Vehicle Technology 2

Vehicle Technology 2

M. J. Nunney, CGIA, MSAE, MIMI

Heinemann : London

William Heinemann Ltd
10 Upper Grosvenor Street, London W1X 9PA

LONDON MELBOURNE JOHANNESBURG AUCKLAND

First published by Butterworth & Co. (Publishers) Ltd 1982
Reprinted by William Heinemann Ltd 1986

British Library Cataloguing in Publication Data

Nunney, M. J.
Vehicle Technology 2
1. Motor Vehicles – Design and construction
I. Title
629.2 TL145

ISBN 0 434 91471 1

Typeset by Tunbridge Wells Typesetting Services Limited
Printed and bound by Whitstable Litho Ltd., Whitstable, Kent

Preface

The encouraging reception accorded to the first level book in this BTEC Series—*Vehicle Technology 1*—would seem to suggest that the particular treatment adopted therein has proven acceptable to BTEC and other students.

This present second level book—*Vehicle Technology 2*—by the same author therefore offers a similar treatment of the material contained in the corresponding BTEC Vehicle Technology 2 Unit .

As before, the opportunity has been taken to amplify the Unit material in respect of historical background, further technical developments and items of topical interest, while still following closely the BTEC system of enumerating the specific objectives. The illustrations have been chosen for their special relevance to the text and apart from those originating from the author, grateful acknowledgement is due not only to the original publishers of this series for allowing the use of illustrations from certain of their other technical books, but also to the following organisations who so kindly supplied the remaining illustrations:

Alfa Romeo (GB) Ltd., Automotive Products Ltd., Bendix Ltd., BL-MG., TI Churchill, Citroën Cars Ltd., Con-Vel Division of Dana Corp., Eaton Ltd., Ferodo Ltd., Fiat Auto (UK) Ltd., Lucas Girling Ltd., GKN Hardy Spicer Ltd., GKN Laycock Engineering Ltd., Goodyear Tyre & Rubber Co., Lada Cars, Lipe-Rollway Ltd., Mercedes-Benz (UK) Ltd., Nissan (UK) Ltd., Pirelli Ltd., Pressed Steel Fisher, RHP Ltd., SAAB (GB) Ltd., Seddon Atkinson Vehicles Ltd., SKF (UK) Ltd. and *Ball Bearing Journal*, Smiths Industries Ltd., Suzuki GB (Cars) Ltd., Alfred Teves GMBH., The Timken Co., Turner-Spicer, Volvo Concessionaires Ltd., Woodhead Manufacturing Co., Yamaha-Mitsui Machinery Sales (UK) Ltd. Illustrations 1.4, 3.18 and 6.30 are reproduced by permission of the manufacturers (c) Talbot Motor Co. Ltd.

M.J.N.

Contents

1 Integral body structure

1.1 INTEGRAL BODY CONSTRUCTION

The integration of frame and body functions into an all-embracing shell is variously known as 'integral', 'unitary', 'chassisless' and originally by the French as 'monocoque' construction. A significant step towards integral construction was the introduction of the 'all-steel' welded body, a concept that was developed in America by the Budd Manufacturing Company for Dodge cars in the late nineteen-twenties. Although a separate chassis frame was still used, it no longer functioned as the total load-carrying member.

Integral construction proper was introduced by Citroën in France during the mid-nineteen-thirties and was evolved in collaboration with the previously-mentioned Budd Company. It was based upon the principle of stressing the outer skin of the body shell to perform a load-carrying function, which contributed to the strength and rigidity of the inner structural members, thereby eliminating the need for a separate chassis frame. Perhaps it may be of interest to mention that the monocoque method of providing a more efficient structure was also well-known to aircraft engineers and featured in such historic World War II aircraft as Blenheims and Spitfires.

Finally we come to modern practice where the principle just described of 'pure' integral construction is often modified in various ways, notably in combining it with separate 'sub-frames' for mounting various parts of the vehicle mechanism. The aim here is chiefly to suppress noise, vibration and harshness from the body interior by the extensive use of rubber isolators and also to simplify the assembly of the mechanical units to the body structure. This particular form of integral or 'semi-integral' construction was pioneered by Daimler-Benz of Germany in the early nineteen-fifties.

Figure 1.1 illustrates the main parts of a modern integral construction body.

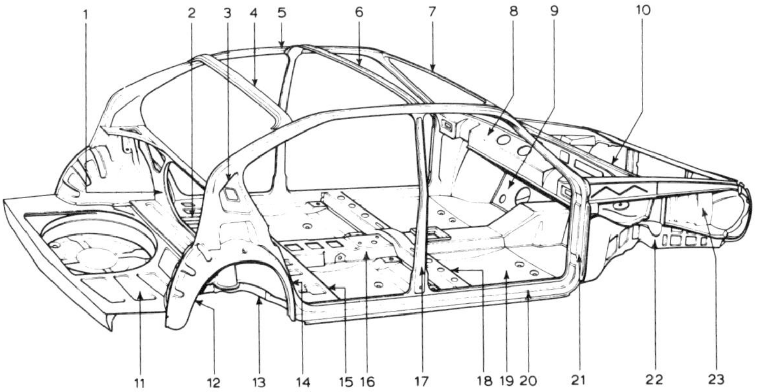

Figure 1.1 Main parts of a modern integral-construction body (*Alfa Romeo*)

1 Rear seat squab bulkhead
2 Rear seat pan
3 Quarter panel
4 Rear window (backlight) reinforcement panel
5 Cantrails
6 Roof bow member
7 Windscreen header panel
8 Dash panel
9 Toe board
10 Service compartment bulkhead
11 Boot floor
12 Wheel arch
13 Longerons
14 'D' post
15 Heel board
16 Central tunnel
17 'B'-'C' post
18 Cross-member
19 Floor panel
20 Sills
21 'A' post
22 Longerons
23 Wing valances

1.1.1 The basic structure

The individual members comprising what is termed the 'body less doors' are most conveniently identified in relation to the following assemblies:

(1) Under-frame.
(2) Side-frames.
(3) Roof-frame.

Under-frame assembly: This may be regarded as the 'backbone' of the modern integral-construction body, since it forms the lower portion of the car and provides the major part of its strength (see *Figure 1.2*). Its middle section comprises a floor ribbed for stiffness

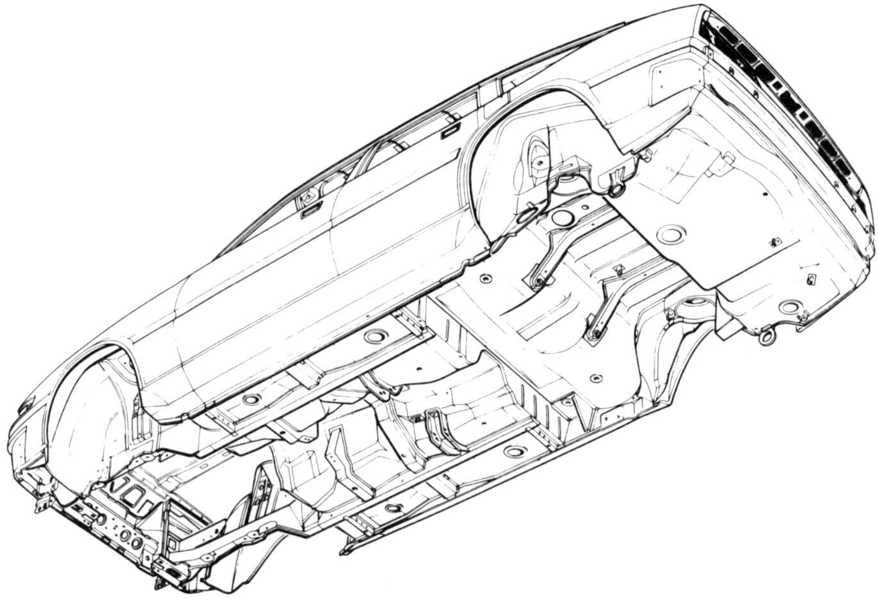

Figure 1.2 Underside view of Rover integral-construction body (*Pressed Steel Fisher*)

and flanked on each side by the inner sill members, and usually it has a central tunnel which houses either the propeller shaft, remote gear control linkage or exhaust system, according to the particular layout of the car. The floor panel terminates at its front end in the toe-board and dash panel sub-assembly and at its rear end in the heel-board and rear seat pan sub-assembly. This typically incorporates the boot floor in one large pressing. Box section cross-members are provided beneath the front seat and at the leading edge of the rear seat pan. A pair of box section 'longitudinal members' or 'longerons', running beneath and either partly or wholly the length of the floor panel inboard of the sills, are extended forwards and rearwards to the body extremities. If the longerons are interrupted along their length, the front pair extend inwards as far as the cross-member beneath the front seats, but the rear pair extend no further inwards than the heel-board cross-member. The under-frame assembly is completed by the front wing valances and wheel arches, which are united with the dash panel and front pair of longerons.
Side-frame assemblies: Each of these comprises an outer sill member, door posts, quarter panel including rear wing, and roof cantrails. With a four-door body the front, middle and rear door posts are

designated 'A', 'B–C' and 'D' posts respectively. In all cases they can pose complex joining conditions with their surrounding structure. Their upper ends are united by what is termed a 'cantrail member', which is also of box-section construction. A side-frame is completed by the rear wing valance and wheel arch, this assembly then being integrated with the rear seat pan and boot floor. Each front wing is attached separately both to its under-frame valance and wheel arch and side-frame door 'A' post.

Roof-frame assembly: This serves to close the essentially box-like structure of the conventional car. It comprises, of course, the roof panel flanked on each side by the cantrails mentioned previously and may be cross-braced by roof bow members. The roof panel joins the windshield header panel at its front end and the reinforcement panel for either a rear window or tailgate at its rear end.

1.1.2 General advantages of integral construction

As compared to the once conventional separate frame and body for car build, the following advantages are usually attributed to integral body construction in its various forms:

(1) Stiffer structure for the same weight.
(2) Improved passenger accommodation.
(3) Safer in a collision.
(4) More economical to manufacture.

Stiffer structure for the same weight: This derives mainly from the elimination of the body-to-frame flexible mountings and provides the opportunity for increasing all-round car performance, without incurring a weight penalty. In particular the body is less prone to shake, which helps to improve both ride and handling qualities. The greater stiffness also contributes to improved durability of the structure, because of its better resistance to the weakening effects of vibration. From this it follows that there should be less tendency for body squeaks and rattles to develop. Another important consideration is that the greater stiffness or rigidity of the structure minimises distortion at the body apertures.

Improved passenger accommodation: The desire to achieve this has always been one of the main reasons for a manufacturer adopting integral construction, and it is gained in two ways. First, the structure is inherently better adapted to the established modern styling requirement for a wide body and hence contributes to greater seating space. Second, it enables the space between minimum ground clearance and the roof to be more efficiently utilised, thereby improving headroom despite a low roof line. This is because the absence of chassis frame side members allows a lower floor relative to the body sills, while the simple cross-members that lend stability to the body floor may be buried under the seats.

Safer in a collision: For some years now increasing attention has been directed by road safety research organisations and manufacturers towards what is called the 'passive safety' of a car (*Figure 1.3*). This term refers to the better protection against injury for the car occupants once an accident has occurred. In fact international legislation is now in force which, among many other passive safety requirements, seeks effectively to limit the amount of deformation that the passenger compartment of a car may suffer in relatively low-speed accident situations.

From our knowledge of basic physics we may recall that the

second of Sir Isaac Newton's three 'Laws of Motion' tells us that when a force acts upon a body, it produces an acceleration which is proportional to the magnitude of the force and inversely proportional to the mass of the body, or in perhaps more familiar terms—'Force equals Mass times Acceleration (or Deceleration)'.

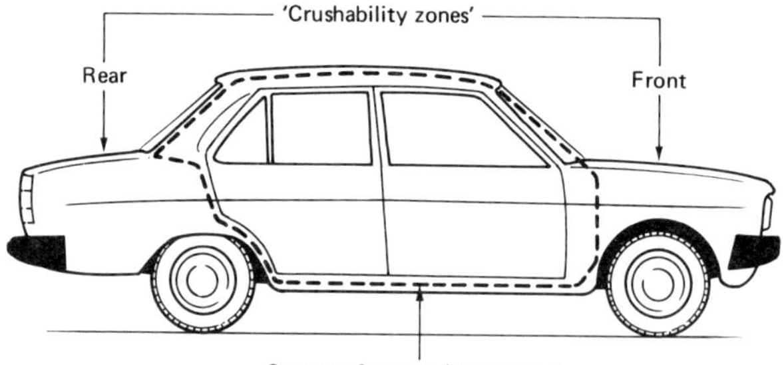

Figure 1.3 Passive safety of integral-construction body

From this it should be evident that it is the rapidity with which the vehicle is stopped that influences the force imposed upon the occupants. Therefore, the very sudden stopping of a vehicle in a crash situation can accelerate the occupants, with respect to the vehicle, such the force of impact may either injure or kill them. In simulated crash situations human volunteers (suitably restrained by energy-absorbing harness) have withstood peak acceleration forces of just over 20g or, stated another way, just over 20 times the gravitational force acting on their own body. On the other hand, it is generally recognised that it would be difficult to sustain life with peak accelerations of 60g or more.

One method of minimising the peak deceleration of a crashing vehicle and therefore the acceleration of its occupants is to absorb the energy of impact over a fractionally longer period of time, by allowing the extremities of the body structure to collapse in a progressive manner. In purely general terms, this can mean that if a vehicle travelling at a speed in the range of 40 to 72 km/h collides head-on with a rigid barrier, then provided that the front end structure of the body can become crushed to a distance of about 0.6 m the peak acceleration suffered by the occupants should not be greater than 40g.

The greater protection afforded by an integral-construction body is achieved mainly by designing the structure so that, unhampered by a separate rigid chassis frame, the front and rear portions of the body may collapse progressively under impact and preserve more nearly intact the middle passenger compartment. In other words, the front and rear portions of the car are deliberately made more 'crushable' than the middle 'cell'. This particular feature of safety body design, involving 'controlled crushability zones' was pioneered by Daimler-Benz in the early nineteen-fifties. Other safety features that may be incorporated in the passenger compartment include a box-section sheet metal 'bar' that spans the rear of the roof panel, which in conjunction with reinforced door pillars provides better protection in 'rollover' accident situations, and to protect against side impacts heavy-gauge steel beams welded inside the doors.

More economical to manufacture: Although the initial tooling cost of integral body construction can reach staggering proportions, the long-term and necessarily high volume manufacturing costs are reduced for several reasons. These include the more efficient use of materials (especially the saving that results from the elimination of a separate chassis frame and its associated body mountings), the reduction in number of parts handled, and the automation of production build techniques.

1.1.3 Less favourable considerations

These are usually concerned with the following aspects of vehicle operation:

(1) Increase in noise transmission.
(2) Earlier deterioration from corrosion.

Increase in noise transmission: The integral-construction body tends to be more affected by panel drumming and road rumble noises, as compared with the separate frame and body. Therefore, resort was made early to such devices as the swaging of floor and other body panels, so as to stiffen them and raise their natural frequency of vibration; the application of sound-deadening treatments on panels and the liberal use of sound-absorbing materials in the offending areas of the body. As mentioned earlier, the semi-integral type of body construction better lends itself to the suppression of noise, vibration and harshness from the body interior.
Earlier deterioration from corrosion: The reason for this is that any corrosive attack on the body panels tends to have a greater weakening effect on the structure as a whole, than would be the case if a separate chassis frame of relatively thicker material was used to support the body. An integral-construction body, produced from pressed sheet metal in thicknesses typically ranging between 0.9 to 1.5 mm according to duty performed, is therefore by its nature more prone to deterioration from corrosion. Protection from corrosion involves, in the first instance, the careful design of structural members to avoid wherever possible joints exposed to the entry of road splash and to ensure that those which are exposed are suitably sealed against it. Secondly, the body structure as a whole must be subjected to selective and thoroughly-applied anti-corrosion chemical treatments. In some body constructions the steel pressings are galvanised in areas especially prone to corrosion.

1.2 STIFFNESS REQUIREMENTS OF INTEGRAL CONSTRUCTION

An integral-construction body must possess considerable stiffness or rigidity for several reasons. The greater its rigidity the higher will be the natural frequency of vibration and hence the lower the amplitude of this vibration. If the former should be too low and unhappily correspond to the frequency of excitation caused by vibration of the unsprung mass of the wheels and suspension, then resonant vibration and body shake can occur. Clearly the body must also be rigid enough to preserve correct alignment of the mounting points either for the front and rear suspension systems direct, or for their intermediate sub-frames. Similarly, the body shell must be able to resist deflection at its apertures for the doors, so that metal-to-metal clearances and optimum sealing conditions are maintained. *Figure 1.4* illustrates various sections contributing to stiffness in an integral construction body.

In resisting the various deflections, the integral-construction body is subject to a combination of loads that may be summarised as being mainly the following:

(1) Vertical bending loads imposed by the gravitational force

acting through the centre of gravity of the car and reacted at the mounting points for the front and rear suspension systems.

(2) Horizontal bending loads resulting from cornering inertia force acting through the centre of gravity of the car and again reacted at the mounting points for the front and rear suspension systems.

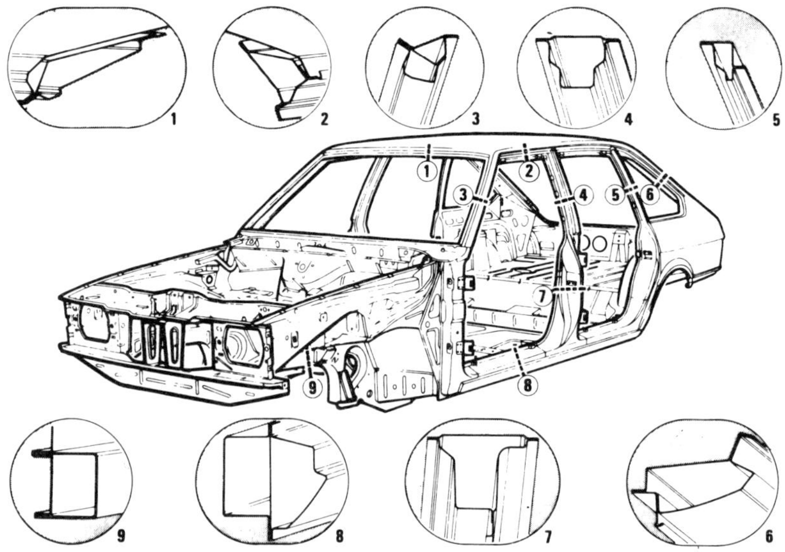

Figure 1.4 Integral-construction body sections for stiffness (*Talbot*)

(3) Torsional loads created by the twisting effect on the body of a single wheel climbing an obstacle and reacted across the mounting points for the front and rear suspension systems. The maximum torsional load condition would occur if the car were supported on only three of its four wheels.

1.2.1 Stiffening effects of various body members

Scuttle: This term has long been used to identify that portion of a car connecting the bonnet and the passenger compartment of the body. Such a clear line of demarcation is now hardly discernible, because the curved and steeply raked windscreen used with modern styling has a lower edge that extends almost to the dash panel. It is, therefore, the latter together with what the Alfa Romeo Company call the 'service compartment' (*Figure 1.5*), which really defines the

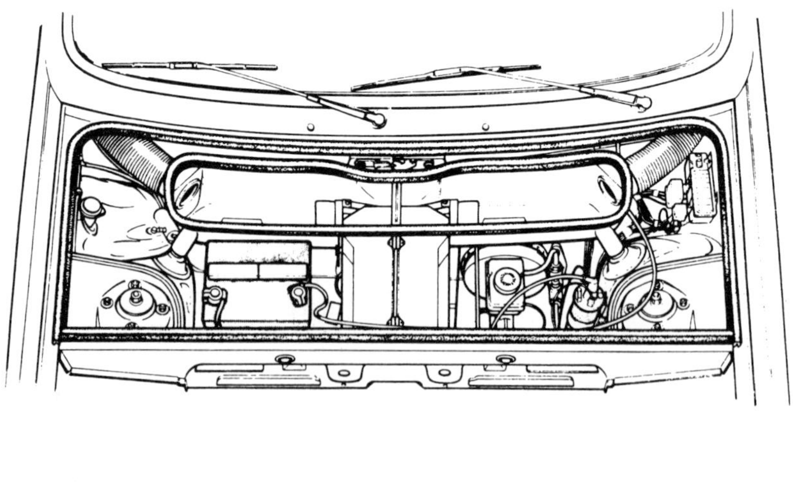

Figure 1.5 Scuttle service compartment (*Alfa Romeo*)

scuttle portion of the modern car. The service compartment forms an enclosed noise-reducing barrier between the engine and the passengers; it separates the outside air used for ventilation and heating from direct contact with the engine; and it contains in one protected area the various service components of the car, such as the battery etc. Structurally, the scuttle portion contributes to the torsional stiffness by acting as a forward cross-tie between the under-frame and the side-frames. A similar role is performed towards the rear of the body by any transverse panel or rear 'bulkhead' that may be used behind the rear seat squab and which spans the rear wing valances.

Underbody assembly: From the description given earlier of the under-frame it should be fairly evident that the longitudinal sills, central tunnel and cross-members promote considerable resistance to bending over the length and width respectively of the floor area. The stiffening effect of the suspension supports, as related to the underbody as a whole, is mentioned later.

Roof bow assembly: In early designs of integral construction the roof bow assembly was generally thicker in section and of greater curvature than is found in modern practice, so that it now contributes rather less to the bending and torsional stiffness of the body. This has been recognised in a recent design where no cross-bracing members are employed for the roof panel, reliance for stiffness being concentrated in the surrounding framework of the roof bow.

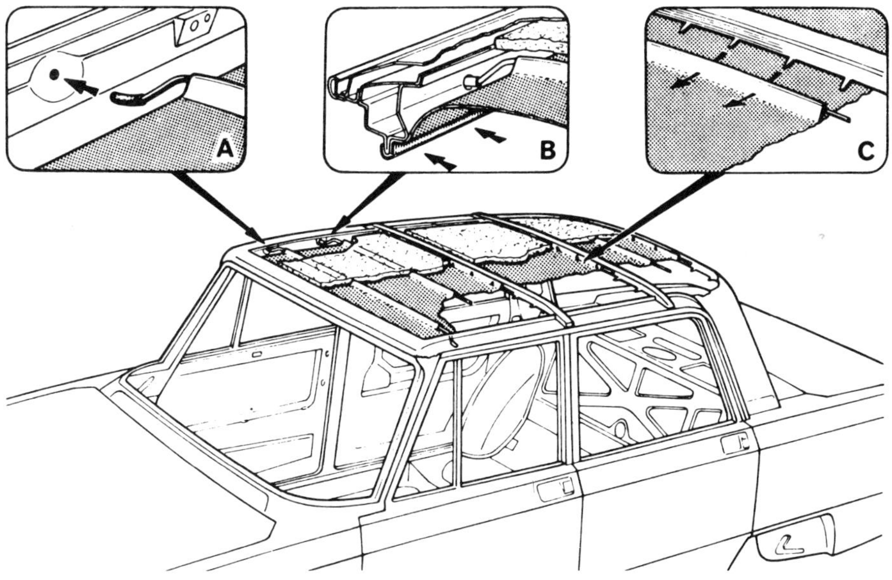

A — Mounting holes for roof trim stiffening bows
B — Areas in which roof trim is stuck to roof panel
C — Hooks for fastening trim bows to roof ribs

Figure 1.6 Details of a roof-bow assembly (*Alfa Romeo*)

Suspension supports: These generally serve to extend the resistance to bending over the entire length of the underbody assembly. Since the longeron members which they normally comprise are integrated with the floor panel and the inswept front and rear wing valances, the supports also contribute to the torsional stiffness of the body.

Bulkhead brace: The term 'bulkhead' originally referred to the vertical flat partition of the scuttle assembly that divided the engine and passenger compartments. In modern practice an often less obvious bulkhead is formed by the complex shaped pressing that comprises the dash panel. A bulkhead brace serves as a rigid triangulating member, both in side and plan views, between the front longerons and the dash panel where it joins the A posts of the side-frames. It may comprise a box-section strut that is disposed diagonally across the sloping inner face of each wing valance, or the bracing effect may be implicit in a sturdy box-like construction for the combination of front wing and valance.
Cantrails: Since the modern styled car has much larger windows or, in the language of the body engineer, 'greater glass areas', there is less space available for structural members in the upper part of the body than was formerly the case. The cantrails (or roof rails) must therefore be of carefully contrived box-sections so that they provide the necessary resistance to bending at the joints between the door pillars and the roof. This bending effect is imposed when, in plan view, the roof tries to rotate relative to the floor as the body is loaded in torsion. That is, the cantrails would attempt to move fore and aft relative to the sills below.

1.3 SUPPORTING FUNCTIONS OF THE BODY

The attachment of the mechanical units to an integral construction body demands an approach different from that adopted for a separate frame and body, because of the relative difference in metal thickness of the load bearing members. More specifically, it is necessary to spread the load to be supported over a greater area of the thinner sheet metal used with integral construction. Otherwise problems could arise in service from localised distortion of the sheet metal sections, which at least could introduce misalignment and at worst result in failure of a particular mounting system. (*Figure 1.7*).

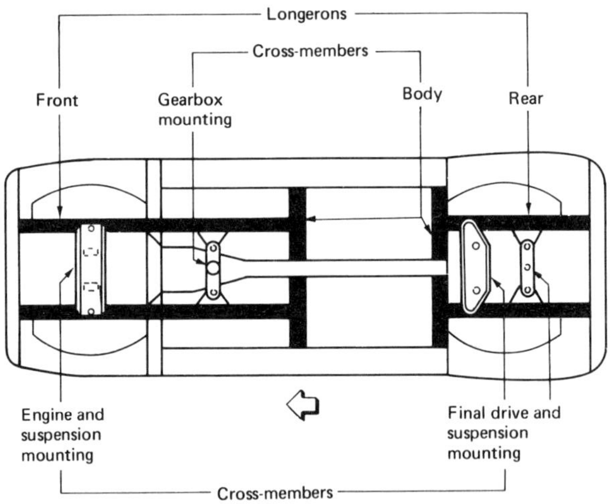

Figure 1.7 Mountings for mechanical units as viewed from underside of integral-construction body

1.3.1 Unit mounting methods

Typical examples of the manner in which the above-mentioned technique is applied are as follows:

Engine and gearbox assembly: With an engine and gearbox unit mounted fore-and-aft, the front engine-mounts are conveniently carried by either a simple cross-member, or a more complex sub-frame, which is primarily used to mount the front suspension mechanism to the body structure. In the former case, a shallow partially-boxed cross-member can be rigidly attached towards the front ends of the longerons.

Alternatively, a sub-frame may be flexibly attached to the longerons at more widely spaced locations along them. In the latter event, the front extremities of the longerons are integrated with a cross-beam to provide the necessary rigid foundation for mounting the sub-frame. The rear gearbox mount is carried by a short cross-member bolted to the underside of the transmission tunnel, this being locally reinforced for the purpose. Less commonly, the engine and gearbox may be carried in their entirety by a flexibly-mounted 'open' sub-frame that extends over the whole length and breadth of the front longerons.

In the case of a transversely-mounted engine and gearbox, the three mounting points are orientated about the unit typically as follows: One mount is positioned beneath the unit at the juncture of the bell-housing and engine block and is supported by a detachable third longeron member running across the bottom of the engine compartment to a front cross-beam. A second mount is positioned at cylinder-head level to support what would normally be the front of the engine, a pick-up point for this mounting being provided on the reinforced valance structure. The third mount takes the form of either a high-mounted stay rod, or a low-mounted torque arm. To control rocking movements of the engine and gearbox unit, the rubber-bushed ends of the stay rod connect to the rear face of the cylinder head and the upper dash panel structure. The alternative torque arm is rigidly attached at its front end to the bell-housing periphery and at its rear end enters a rubber bushing supported from the lower dash panel structure. With both arrangements the mounting support areas are locally reinforced on the dash panel assembly. An alternative method of mounting a transverse engine and gearbox is to employ a four-point system, as found in the early Metro for example, where the forward and rearward pairs of mounts are carried on cross-members from the longerons, the rearward pair inclined towards the unit for better torque control.

Final drive: If the rear wheels are the driven pair and are also independently sprung, it is necessary to provide mounting points for the final drive unit on the car body structure. In conventional practice, however, the final-drive unit is supported either wholly, or partly, by the flexibly-mounted sub-frame from which the suspension arms are pivoted. The sub-frame is attached beneath the forward ends of the rear longerons. Where an additional mounting is required for the rear end of the final drive unit, it is usually supported from a stiffening cross-beam integrated with the boot floor and the longerons.

Front suspension: As mentioned previously in connection with engine mounting, the front suspension mechanism is supported from the body structure either directly by means of a rigidly attached and simple cross-member (*Figure 1.8(a)*), or indirectly through the medium of a flexibly-mounted sub-frame.

We must now qualify this statement by observing that modern IFS systems of the MacPherson strut-and-link type require additional supporting features. In particular, a pair of forward anchorage

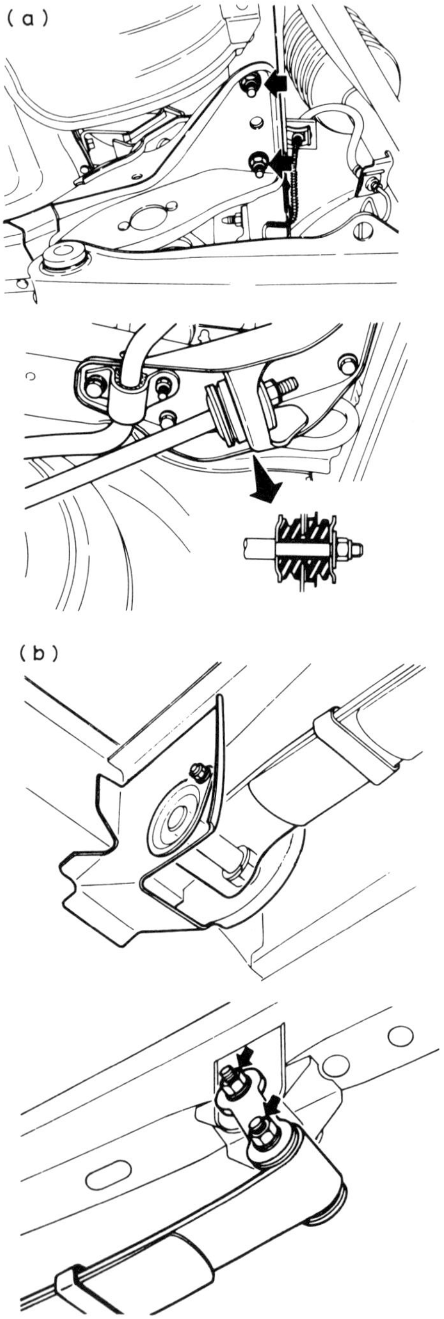

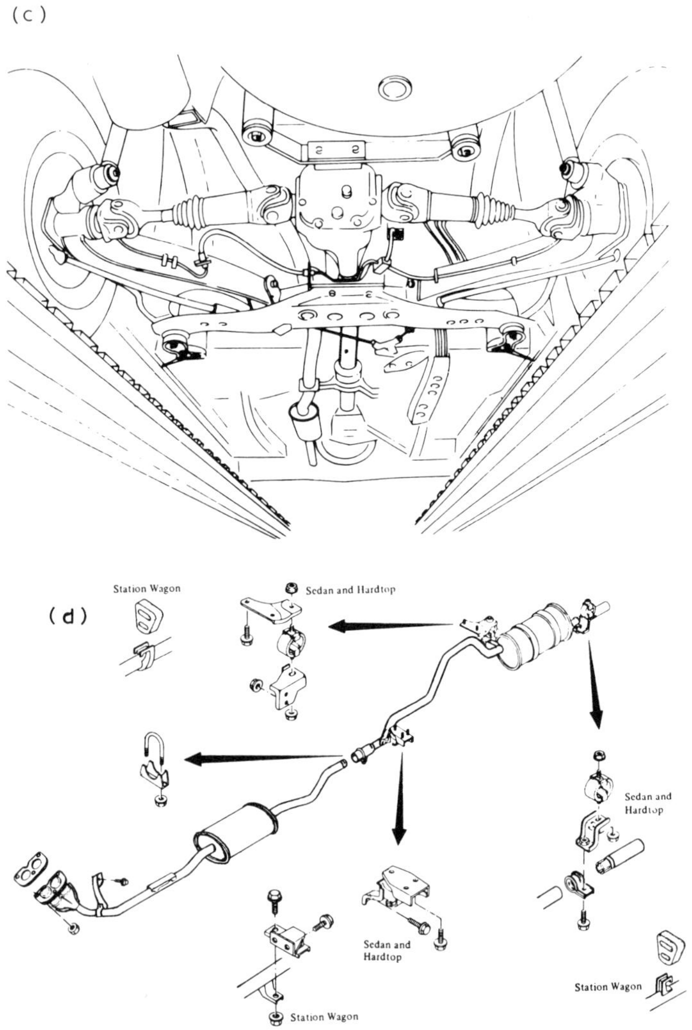

Figure 1.8 Unit mounting methods for integral-construction body. (a) Cross-member and mounting brackets for engine and front suspension, (b) mounting brackets for rear suspension semi-elliptic leaf springs, (c) cross-member and mounting brackets for IRS, (d) mounting brackets for exhaust system (*Nissan*)

brackets are required where diagonal links are used to transmit fore-and-aft forces from the track control arms to the body structure, the arrangement being described in § 6.4.2 of *Vehicle Technology 1*. With this type of system an upper flexible mounting must also be provided for each strut, and since the suspension spring load is reacted at the same point, a very rigid and suitably-braced 'MacPherson tower' is an essential feature of the wing valance construction. The widely-spaced attachment points for this type of suspension system do, nevertheless, make a valuable contribution to reducing load concentration on the body front end structure, as indeed was the original design intention.

Rear suspension: In conventional practice the suspension mounting for driven rear wheels, independently sprung by a semi-trailing arm system, is effected through the medium of a sub-frame flexibly mounted towards the forward ends of the rear longerons. An upper location for each suspension coil-spring and shock-damper is provided against the upswept portion of the longerons, which may also be locally increased in width to accommodate them.

In contrast, the increasingly popular method of suspending non-driven rear wheels by means of a modified strut-and-link system, requires a substantial box-section cross-member from which the transverse links can pivot and against which the suspension springs can react. It is rigidly attached towards the rear of the body sills, which in this case are extended under the boot floor to furnish a wide base mounting for the rear suspension system (*Figure 6.10*).

For live-axle rear suspension systems the pivot points for flexibly mounting either semi-elliptic leaf springs (*Figure 1.8(b)*), or positive locating links as described in § 6.17, are provided by external brackets and internal sleeves integrated with the rear longerons.

Exhaust system: The installation of the flexibly-mounted exhaust system to the under-frame of an integral-construction body, varies in detail according to the geographical layout of the system. For a directly-routed single system, the front and the rear silencer with tailpipe are supported from mounts that are typically located from rigid body sections towards the centre of the rear seat pan cross-member and the far end of the appropriate rear longeron. With a longer wheelbase car an additional forward support for the system, taken say from the cross-member under the rear gearbox mount, is also likely to be employed. If any form of dual exhaust system is used then there must, of course, be a duplication of the mounting arrangements for at least part and in some cases the whole length of the system. The mounting brackets required for an exhaust system are shown in *Figure 1.8(d)*.

1.4 ANTI-VIBRATION MOUNTINGS

By the very nature of its rotating, reciprocating and oscillating mechanisms, the running of a car is attended by a certain amount of vibration. Even though such vibration is minimised at source in modern automotive practice, its transmission to the car body interior either simply as vibration or in the form of noise audible to the occupants, or both, is always undesirable. Indeed, its suppression is the declared mission in life of the 'Noise, Vibration and Harshness' (NVH) test engineer, a specialist activity first identified by the Ford Motor Company in the early nineteen-seventies when increasing attention began to be given to this problem.

1.4.1 Methods of reducing NVH

The ideal, but albeit totally unrealistic, method of confining vibration to any particular mechanical unit would be to 'suspend it in space', so that it could vibrate without causing a disturbance in the car. In reality, the next best thing is to support the unit on flexible 'anti-vibration mountings' in which rubber is generally employed as the spring medium. By using a flexible, as opposed to a rigid, mounting system the vibratory forces of a mechanical unit are reduced to the relatively small spring forces transmitted by the support mountings themselves.

At this point, we should perhaps distinguish between vibration 'isolation' and 'insulation'. For example, the engine rubber mounts not only isolate the car structure from outgoing vibrations of the engine, but also insulate the engine against incoming vibrations from the car structure, so that the overall effect is for vibration of the engine unit to be minimised under all running conditions.

An advantage of using rubber, rather than steel, as the spring medium for anti-vibration mountings is that the sound transmission through them is reduced. Basically this can be attributed to the absence of a metal-to-metal path for the sound to travel along, since rubber in itself is not particularly sound-absorbent owing to its natural frequency being small compared with that of sound. Another advantage is that the internal friction of rubber tends to dampen vibration. However, it is worth reflecting that no matter how much a car may cost, its acceptable functioning depends very much on the few 'blocks of rubber' used to support its mechanical units. The various applications of anti-vibration rubber mountings can be readily identified from the illustrations in *Figure 1.9* showing front and rear suspension systems together with their sub-frames.

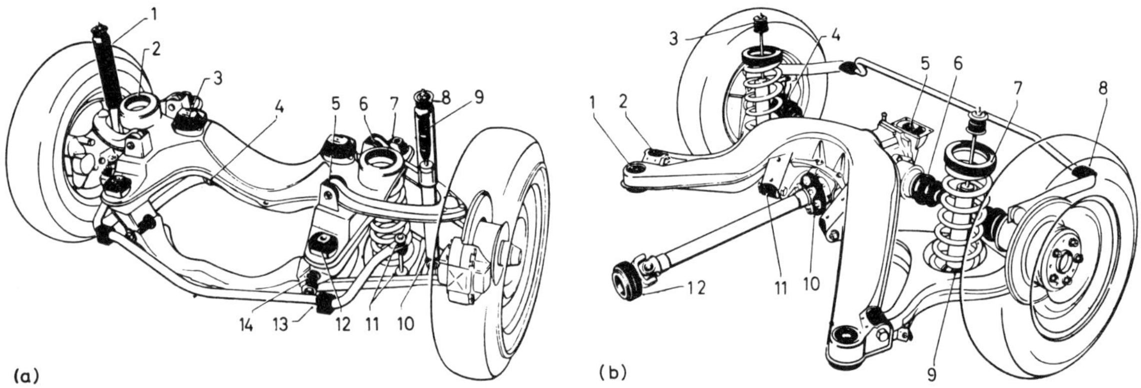

(a) *Front axle:* 1, protecting sleeve; 2, ring; 3, engine mounting; 4, bush of the lower lateral wheel fork arm; 5/6, bush of the upper transverse wheel fork arm; 7, front axle carrier mounting; 8, shock-damper rings; 9, spring washers; 10, shock-damper mounting; 11, pad; 12 front axle carrier mounting; 13, torsion bar bearing; 14, bush of the lower lateral wheel fork arm
(b) *Rear axle:* 1, rear axle carrier mounting; 2, diagonal wheel fork arm bush; 3, shock-damper pad; 4/6, half-axle gaiters; 5, rear axle centre mounting; 7, coil spring mounting; 8, torsion bar bearing bush; 9, shock-damper mounting; 10, prop shaft coupling plate; 11, diagonal wheel fork arm bush; 12, prop shaft bearing
(c) *Front axle carrier detail:* 1, body; 2, mounting; 3, flange plate; 4, bumper pad; 5, front axle carrier

Figure 1.9 Reducing noise, vibration and harshness by the use of rubber parts on axles (*Mercedes-Benz*)

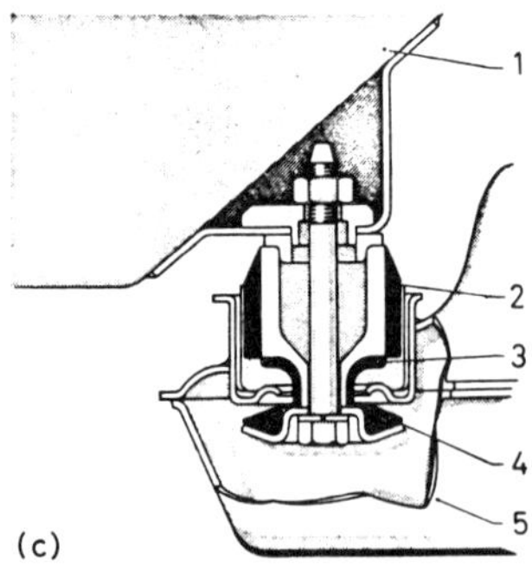

All the thin metal panels of an integral-construction body likely to vibrate or transmit noise are covered with anti-vibration materials of one type or another. They generally take the form of 'sound deadening' treatments, such as the application under heat of bituminised material direct onto the panels; and 'sound absorbing' materials in pad form, such as resin-bonded layers of

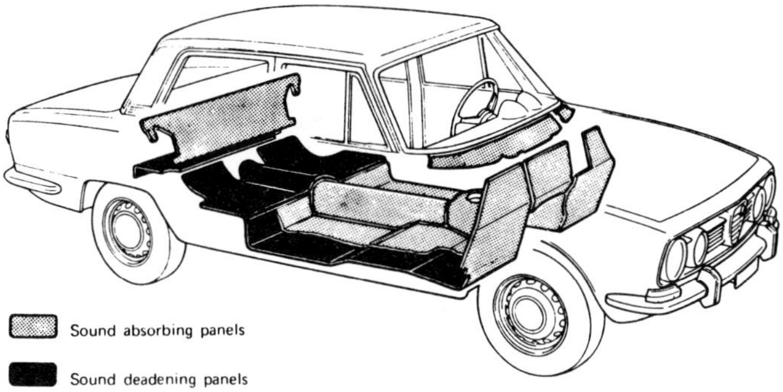

Figure 1.10 Sound-proofing of integral-construction body (*Alfa Romeo*)

natural and artificial fibres that are cemented to the panels (*Figure 1.10*). Also, the various box-section stiffening members of the body may be filled with foam products. In all cases the principle is the same—the greater the difference in 'sound resistance', as defined by the product of sound velocity and material density, between the materials in contact, the more effective is the sound damping.

2 Friction-clutch mechanisms

2.1 CLUTCH CENTRE-PLATE CONSTRUCTION

In § 2.2.1 of *Vehicle Technology 1* the centre-plate was identified as the driven member of the 'plate' type friction clutch, because it is sandwiched between the friction contact surfaces of the flywheel and the clutch pressure-plate. At this stage we must consider the construction of the centre-plate in a little more detail, especially those features that contribute to the overall smoothness of torque application and transmission. In this context the modern clutch centre-plate usually incorporates at least an 'axial cushioning' facility between the friction linings and, except in arduous duty applications, also a 'torsional cushioning' facility between the plate and its hub. It is therefore convenient to classify the two types of centre-plate construction as follows:

(1) Solid drive.
(2) Cushion drive.

Solid drive: This does, of course, represent the simplest form of centre-plate construction. The hub and the plate carrying the pair of friction linings are directly riveted together and thus provide a torsionally rigid drive through the clutch assembly. (*Figure 2.1*).

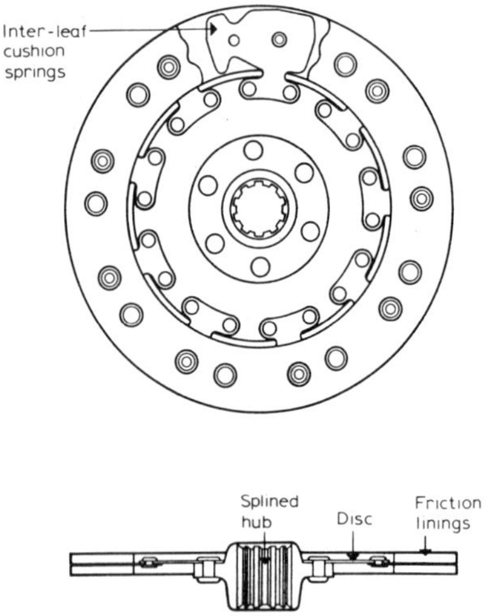

Figure 2.1 Clutch centre-plate with solid drive (*GKN Laycock*)

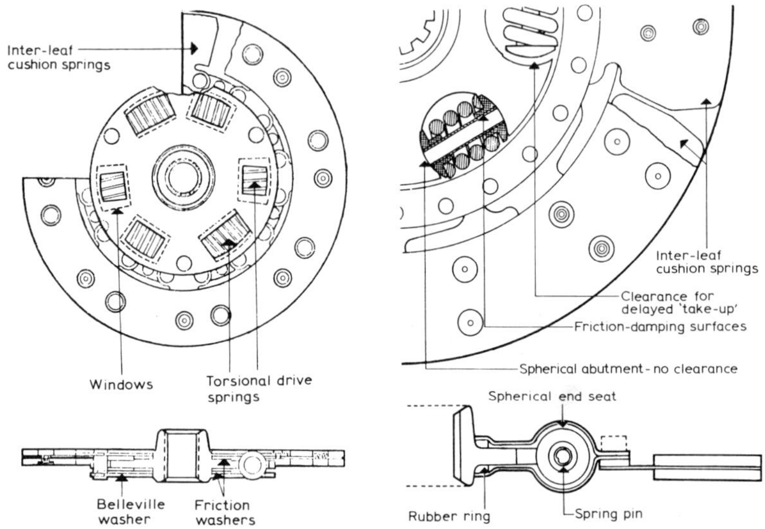

Figure 2.2 Two types of friction-damper for cushion drive (*GKN Laycock*)

Cushion drive: In this construction the plate carrying the pair of linings is provided with a complementary side-plate to form a sandwich with the flanged hub, so although the plate assembly is positively located both axially and radially, in the rotational sense its location is effected through a number of circumferentially-spaced compression springs. These are retained in 'windows' formed in the hub and plate sandwich. This feature thus allows a limited degree of angular wind-up of the centre-plate about its hub, or in other words torsional cushioning. (*Figure 2.2*).

2.2 FUNCTIONING OF THE CENTRE-PLATE COMPONENTS

Hub: This is provided with a spline connection to the primary shaft of the gearbox. A sliding fit is permitted so that when the clutch is disengaged, the friction-lined centre-plate can float between the separated contact surfaces of the flywheel and the pressure-plate. Either 'square-sided' or 'involute-form' splines may be specified for this application. An involute spline has teeth similar in form to those of a gear and since this topic receives attention in § 3.2.1, suffice it to say here that the advantages of the involute over the square-sided spline are generally those of possessing greater strength and endurance, a better self-centring of the parts, and more economical to produce.

Torsional spring drive: This feature is incorporated between the centre-plate and its hub for the following reasons:

(i) To minimise passenger discomfort and reduce shock loading on the transmission components in the event of too sudden engagement of the clutch. This is achieved by allowing a controlled degree of wind-up between the centre-plate and hub, the circumferential springs momentarily acting as energy storage devices.
(ii) Fluctuations in driving torque are inherent to the operation of the reciprocating piston engine. If they are passed on undiminished by the clutch to the vehicle transmission system, then vibration and noise can result from any source of clearance therein such as splines, dog-clutches, synchroniser assemblies and gear teeth. In this case the spring drive is required to act as a 'torsional vibration damper', which not only involves the previously-mentioned wind-up between the centre-plate and hub, but also the introduction of a certain amount of frictional resistance to this movement. This is to dissipate the energy of vibration in the form of heat.

The manner in which this feature is built into the centre-plate has already been described in part in § 2.1. So far as the cushioning springs are concerned, it can be added that a greater degree of wind-up tends to be demanded of the modern clutch centre-plate. This may entail using springs of different rates (the 'rate' of a spring being the load required to produce unit deflection). Their purpose is to control not only the higher magnitude vibrations from the engine, but also lower magnitude vibrations from the transmission drive-line. Limiting stops are provided to prevent the springs from being compressed choc-a-bloc (coils in contact with one another).

Finally, the earlier-referred to frictional resistance or 'damping' of centre-plate wind-up and recovery motions may be accomplished in two ways, as below:

(i) By friction washers positioned on both sides of, and spring-loaded against, the flanged hub of the centre-plate. The spring loading may be supplied by a Belleville spring, the principle of which was described in § 2.2.2 of *Vehicle Technology 1* in connection with the clutch pressure-plate.
(ii) By mounting the torsional drive springs between end caps that

slide along specially-formed pins, the friction occurring between these parts supplying the necessary damping action.

Dishing and slits: The dishing of a centre-plate is concerned with its axial cushioning facility, (*Figure 2.3*), which is provided for the following reasons:

(i) In combination with a suitable choice of liner material, the axial cushioning should permit a smooth, gradual clutch engagement, free from any judder that may be inherent to the vehicle transmission arrangements.

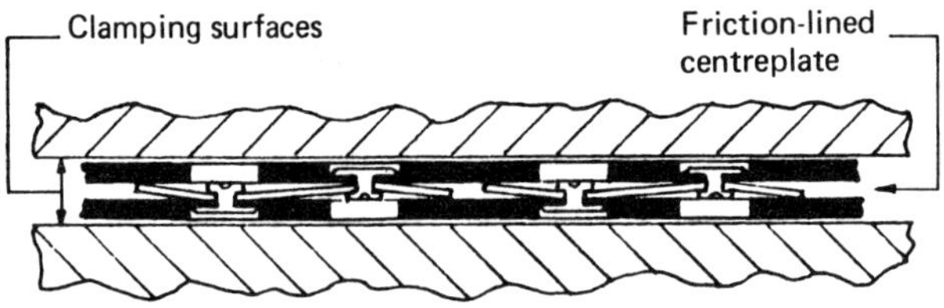

Figure 2.3 Cross-section through axially-cushioned friction linings

(ii) It reduces the initial pedal load required to disengage the clutch, because the spring force due to the axial cushioning opposes that exerted on the liners by the clutch pressure plate.
(iii) The clutch pedal action or 'feel' is improved by virtue of the engagement process occupying a greater pedal travel, so that a more gradual transition from slip to grip can be made. In some early clutch installations where the pedal travel was very short, the all-or-nothing pedal action could often result in the once well-known 'kangaroo start'!
(iv) Longer wear life of the friction liners may be obtained due to their more satisfactory bedding with the friction contact surfaces of the flywheel and the clutch pressure-plate, thereby allowing for a uniform heat generation over the entire friction surfaces. Wear is always more rapid if localised hot-spots occur on friction surfaces with uneven bedding.

The axial cushioning is provided through some form of dished spring arrangement. For this purpose the friction linings are attached to the centre-plate either directly, in which case the outer portion of the plate is provided with a series of crimped tongues, or indirectly through the medium of spring-steel, axially-curved segments that are separately riveted around a reduced size of plate. The latter method has long since been preferred because it avoids any tendency for the liners themselves to become dished. A parallel approach of the friction contact surfaces is generally essential for judder-free engagement of the clutch.

With regard to centre-plate slits, the stresses that are set up in a thin metal disc when heated are such that it tends to distort in a wavy fashion. To relieve a disc of these stresses it is customarily provided with slits or 'radial expansion slots', which extend to the circumference and terminate at their inner ends in small circular holes, again to avoid stress concentration. It is clearly desirable that the centre-plate should remain free from distortion if it is not to drag when the clutch is disengaged.

Friction linings: For the clutch centre-plate friction linings either asbestos-based or, less commonly, sintered metal materials are used. The single most important difference between these materials lies in their relative thermal conductivities, which are between 10 and 20 times as great for the sintered metal. This material is therefore generally confined in application to heavy-duty commercial vehicle clutches where it can provide cooler operation and longer life. Friction lining materials and properties will be further reviewed in § 2.3.

Although clutch friction linings are normally attached to the centre-plate with rivets, bonding with a thermosetting adhesive may instead be adopted. There are several technical advantages associated with bonded linings, including an increased effective life because of the greater volume of friction material available for wear, and less possibility of scoring due to the absence of any accumulations of abrasive particles. From practical considerations, however, the greater ease of friction lining replacement made possible with the riveting method continues to guarantee its popularity.

2.3 CLUTCH FRICTION LINING MATERIALS

It is recorded in a history of the Ferodo Company that when London motor buses first came on the roads fitted with leather clutch linings, these sometimes had to be replaced twice in a day because the drivers had a habit of slipping them and burning them out. It was also not unknown for drivers of fairly powerful cars to have the clutch relined en route on journeys of, say, 150 miles; and racing cars almost always had to be relined after a speed test or sporting event. In this field Herbert Frood, who founded the Ferodo Company ('Ferodo' is an anagram of Frood with one letter added), and one or two colleagues first carried out tests on their own cars by replacing leather with impregnated cotton for the clutch linings. As motoring enthusiasts they were very pleased with the results, which gave long wear, steady friction and smooth engagement. The first known use of impregnated cotton clutch linings on public service vehicles took place about 1905, when they were adopted by the Scottish Motor Buses of Edinburgh and quickly followed by the London General Omnibus Company.

An important point in the development of motor-racing in Britain and the British car generally was the opening of the Brooklands racing track in 1907. In those days racing cars gave their constructors the opportunity to test in practice new elements of design, since the conditions experienced in racing accentuated many problems likely to be encountered with standard production cars, although this is less true of racing now. One such problem was bringing into prominence the need for clutch and, of course, brake lining materials with greater resistance to heat and wear than were hitherto available. This led to Herbert Frood experimenting with woven asbestos fabric as a possible basis for high-temperature friction linings. The tests he conducted on this material showed that a combination of pure woven asbestos spun on brass wire to increase its strength, together with a new bonding agent he had developed, produced a friction material possessing the high temperature-resisting qualities he was seeking. One of the first users of the new material—Ferodo bonded asbestos, or BA as it was later known—was once again the London General Omnibus Company.

However, there is now a trend away from what became the highly developed asbestos-based friction liners that we have known for so many years, in favour of environmentally safer asbestos-free materials.

2.3.1 A typical clutch-lining material

General description of Ferodo 2124F: This is a rigid, moulded friction material, grey brown in colour, having a random fibre asbestos base and containing brass chippings. Although intended for use under dry operating conditions only, the material is physically unaffected by oil contamination. The material has high friction with good resistance to fade and particularly good resistance to wear, especially under heavy-duty conditions; it has excellent take-up characteristics and lacks any tendency to score the mating surface. For design purposes its coefficient of friction μ is quoted as 0.38. Perhaps it should be explained here that the 'coefficient of friction'—denoted by the Greek symbol μ (mu)—is the ratio of the force F overcoming friction to the normal load W acting between the sliding surfaces, so that we can express this simply as $\mu = F/W$.

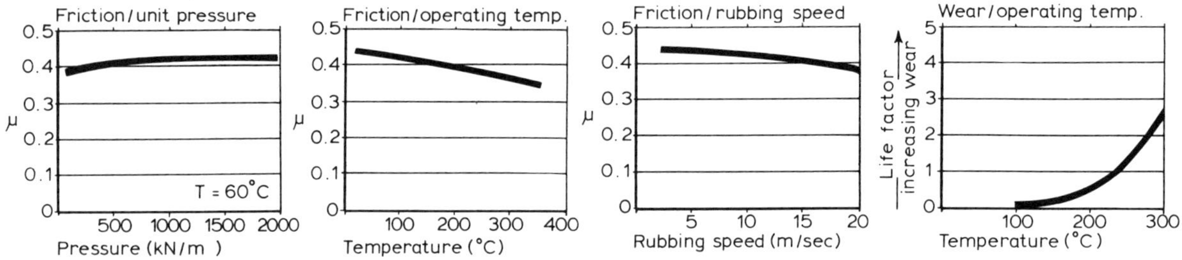

Figure 2.4 Operating characteristics of Ferodo 2124F clutch-lining material

The operating characteristics of Ferodo 2124F are illustrated graphically in *Figure 2.4.*

2.3.2 Sintered metal clutch linings

As mentioned previously, sintered clutch linings or segments are generally confined to heavy-duty commercial vehicle applications. However, a brief note on this type of friction material may be of interest. The process of sintering may be regarded as the squeezing together and heating of very fine particles of metal, such that very strong adhesion occurs between them. In the case of commercial vehicle clutches a bronze-based sintered metal friction material may be used. Although such materials are initially more expensive than the traditional asbestos-based ones, the advantages to be derived from their use are generally as follows:

(1) Cooler operation and a longer life as a result of their much higher thermal conductivity.
(2) More stable friction characteristics under difficult operating conditions.
(3) Can withstand high unit pressures without their performance being affected.
(4) Low rate of wear reduces the frequency of manual clutch adjustment.
(5) Thinner linings (or segments) can be used to provide a more compact clutch unit, especially if it is of the multi-plate type.

2.4 CLUTCH-RELEASE BEARINGS

The function of a clutch release bearing was made clear in § 2.3 of *Vehicle Technology 1.* Its construction may be considered under two headings, as below:

(1) Metallised carbon ring.
(2) Ball-bearing race.

2.4.1 Metallised carbon ring

This type of clutch release bearing (*Figure 2.5*) is of the 'plain thrust' variety and therefore comprises two members. First, there is the rotating member of 'counterface', this being an annular steel plate carried on the tips of either the release levers or the diaphragm fingers of the pressure-plate assembly. Second, there is the non-rotating member that takes the form of a metallised carbon ring, protruding from a cast-iron carrier in which it is rigidly held. The pivoting carrier is in turn mounted on a forked withdrawal lever and retained by a pair of spring clips.

Although this type of release bearing is by the nature of its material inherently self-lubricating, the life that can be expected from it is limited—as in the case of other dry rubbing bearings—by allowable wear. However, the wearing of this type of release bearing did tend to compensate for wearing of the clutch friction linings, since the effect of the latter was to advance the counterface towards the ring. Nevertheless, by the late nineteen-sixties it began to be found that the life of the dry rubbing release bearing could become unacceptably short in constant heavy traffic driving conditions and this led to it being superceded by the ball-bearing type.

2.4.2 Ball-bearing race

A release bearing of this type must be provided with an axial, rather than an arcuate, withdrawal movement. This is because its rotatable element directly engages either the release levers, or the diaphragm fingers, of the clutch pressure-plate assembly. For this reason the bearing must be sleeve-mounted and guided along a nose-piece extension of the cover for the gearbox primary shaft bearing. The release bearing sleeve must also be provided with either a collar, or a recess, to intercept the withdrawal fork of the clutch control system.

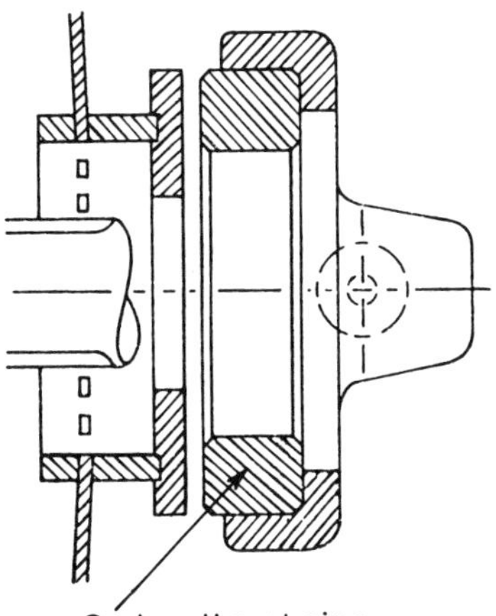

Figure 2.5 Clutch-release bearing of carbon-ring type (*GKN Laycock*)

The modern clutch-release bearing of the ball-race type has passed through several phases of development. Originally a pure thrust bearing was used, which could be likened to a pair of flat washers separated by a ring of ball bearings, their circular path being maintained by matching grooves in the opposing faces of the washers. It was provided with its own shroud and greased for life during manufacture. A drawback with this particular type of release bearing was that its operation at high engine speeds could give rise to destructive skidding between the ball bearings and their raceways due to centrifugal effects.

A single-row, deep-groove, type of ball-bearing race was the next choice and is still retained by some manufacturers, especially for heavy-duty clutch release bearings. This type of ball bearing, although having a high radial load rating, also has a substantial thrust resistance, which is of course the relevant factor. It further possesses the well-known features of being simple in design, non-separable and suitable for high-speed operation. Again the bearing is provided with its own shroud or housing and is often greased for life during manufacture, provision for re-greasing in service being made in some heavy-duty applications (*Figure 2.15*).

Finally, we come to the modern single-row, angular-contact, ball-bearing race. This type of release bearing is distinguished by having a high shoulder on one side of the outer ring and usually another on the diagonally opposite side of the inner ring. The relative positions and steep contact angle of these shoulders give the bearing a high thrust resistance in the required direction of clutch withdrawal. As an alternative to the traditionally machined bearing rings, more recent

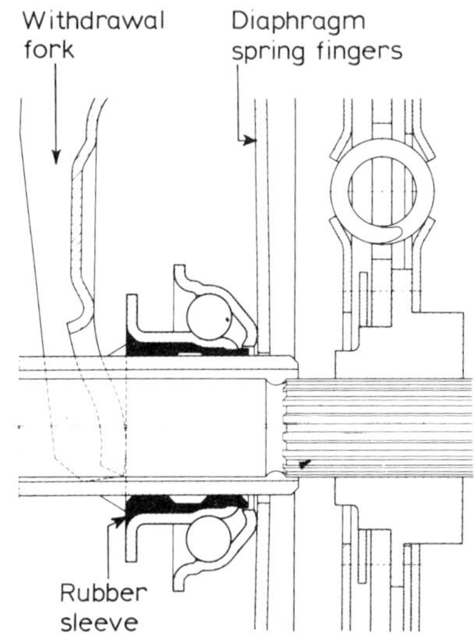

Figure 2.6 A modern self-aligning clutch-release bearing of angular-contact ball type (*SKF*)

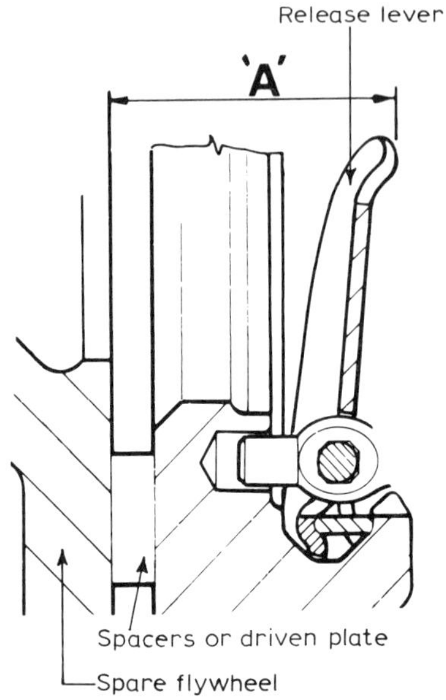

Figure 2.7 Release levers setting for coil-spring clutch. A — height of levers (*Seddon Atkinson*)

versions may be produced from pressed-steel metal rings and sliding sleeves. The form of the bearing rings imparts the necessary rigidity and also contributes to the permanent sealing arrangements. Since the points of contact between the release bearing and the diaphragm spring fingers must alter as the latter are deflected, it is necessary for either a convex or a flat-faced bearing to be used in combination with flat and curved spring finger ends respectively. To improve the durability of the first arrangement, a recent innovation by the SKF Company is to make the release bearing self-aligning through a rubber sleeve in the bore (*Figure 2.6*). Hence, any slight eccentricity of the release bearing mounting relative to the rotating clutch assembly can be accommodated, thereby avoiding excessive side-loading and wearing of the diaphragm spring fingers.

2.5 RELEASE LEVERS SETTING FOR COIL-SPRING CLUTCHES

The satisfactory operation of the multi-coil-spring type of clutch, as described in § 2.2.1 of *Vehicle Technology 1,* is very dependent upon the accuracy of the release levers adjustment. From what has previously been said in relation to the parallelism of the centre-plate friction linings, it should likewise be evident that the pressure-plate must always move parallel with the flywheel. In practice, this means that all three or more release levers must be adjusted to the same distance from the flywheel friction-face, a uniformity within 0.12 mm being typically required. This is achieved by turning the nut on each eyebolt to adjust the height of the levers. When the correct lever setting is obtained, each eyebolt adjusting nut is staked into the slots in the eyebolt. The adjustment is required, of course, only during manufacture or rebuilding of the clutch in service. (*Figure 2.7*).

2.5.1 Spring fingers setting for diaphragm clutches

A manufacturer may also require that the 'toe height' of the diaphragm spring fingers be checked and if necessary adjusted. The

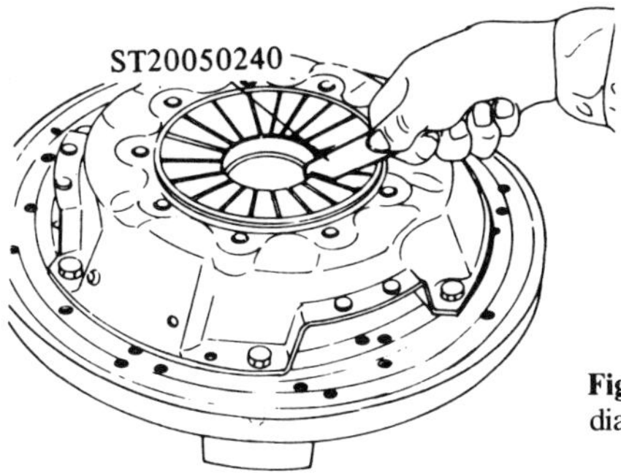

Figure 2.8 Spring fingers setting for diaphragm clutch (*Nissan*)

toe height is checked in a similar manner to that already described for the release levers of coil-spring clutches, but adjustment is performed by using a special setting tool on the spring fingers. Any unevenness of toe height should typically be kept within less than 0.5 mm. (*Figure 2.8*).

2.6 CLUTCH PEDAL AND LINKAGE ADJUSTMENT

When the clutch is engaged the clutch pedal must be retained at a certain distance above the toe-board, this distance being known as the 'pedal height'. It allows the pedal sufficient movement to ensure that the clutch can be completely disengaged, but not too much movement that would cause over-travel and, by the same token, over-stressing of the clutch release mechanism.

There is also another factor to be taken into account. With gradual wearing of the centre-plate friction liners, the clutch pressure plate assumes positions closer to the flywheel. This results in an opposite, but albeit magnified (due to leverage), movement of the inner ends of either the release levers or the spring fingers towards the release bearing. Unless this movement can take place unimpeded by the clutch control system, then clearly the effect would be similar to that of the driver gradually disengaging the clutch—or in other words, the clutch will begin to slip.

In modern transmission practice, there are two basic approaches to accommodating this unwanted movement at the clutch release bearing, these being conveniently classified as follows:

(1) Non-continuous contact.
(2) Continuous contact.

2.6.1 Non-continuous-contact system

The essential features of both mechanical and hydraulic versions of this once traditional method of accommodating the effects of lining wear are that the clutch pedal height remains constant, but included in the pedal movement is a certain amount of 'free-play' (*Figure 2.9*). The latter is to ensure that a definite clearance exists between the clutch release bearing and either the release levers or the spring fingers, at all times when the clutch is engaged. Hence, the levers or the fingers are permitted a limited amount of unrestricted movement towards the release bearing in its 'off' or no contact position (*Figure 2.10*).

Wearing of the clutch liners is, or at least should be, signalled to the driver by the decreasing free-play felt at the clutch pedal. A screw thread adjustment is provided in the clutch control system to restore the free-play, which generally lies in the range of 20 to 30 mm and is checked in service every 6000 miles (10 000 km).

In the case of a cable-operated system, a screwed sleeve with

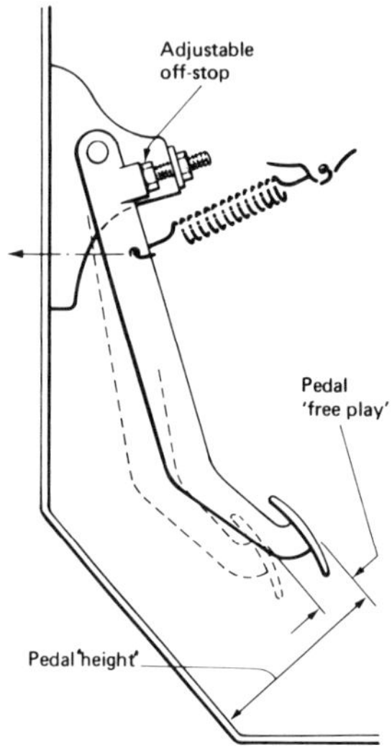

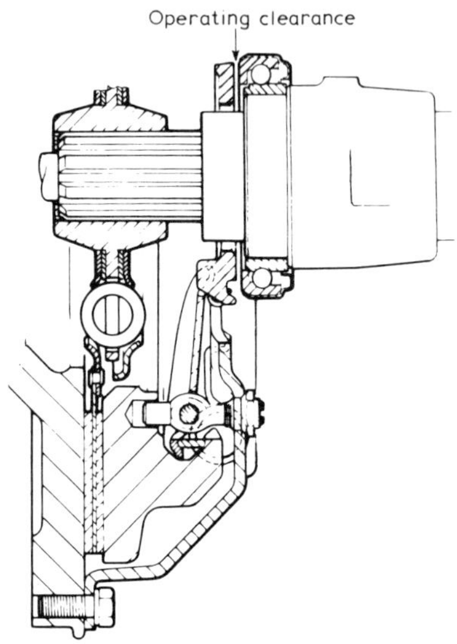

Figure 2.9 Clutch-pedal adjustment, non-continuous-contact system

Figure 2.10 A non-continuous-contact clutch release system (*Seddon Atkinson*)

adjusting nuts is incorporated at one end of the outer cable, usually where it is anchored to the bell-housing. For a hydraulically-operated system, the screw and adjusting nuts facility is provided on the end of the slave cylinder push-rod, where it connects to the clutch release fork (*Figures 2.11* and *2.12* of *Vehicle Technology 1*).

2.6.2 Continuous-contact system

Clutch control systems of this type are of more recent origin and involve continuous contact of the release bearing with the spring

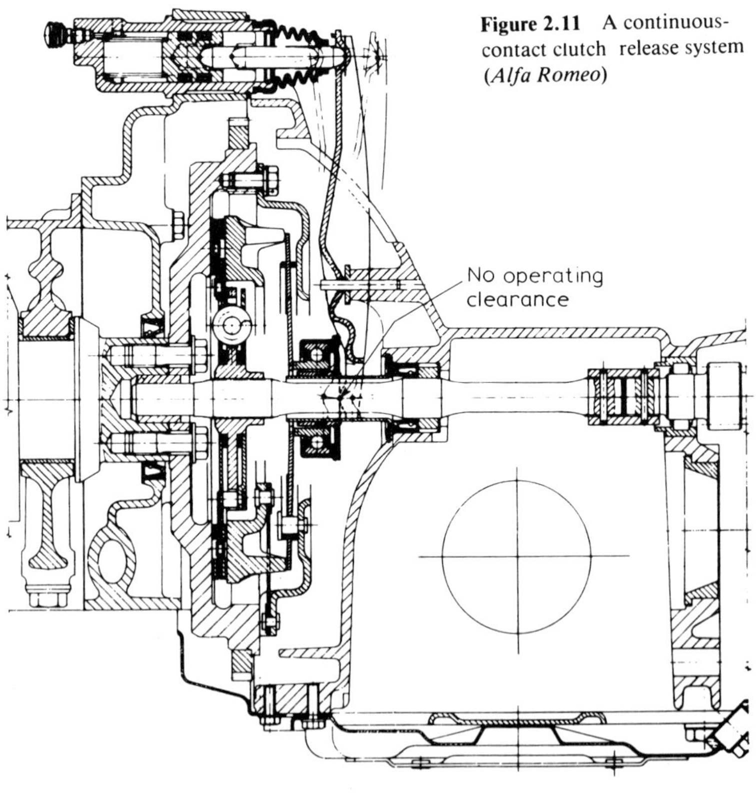

Figure 2.11 A continuous-contact clutch release system (*Alfa Romeo*)

fingers of the clutch unit, regardless of whether or not the clutch is disengaged. To all intents and purposes such systems are self-adjusting to compensate for wearing of the clutch linings. The manner in which the release bearing is permitted to retreat in sympathy with unwanted movement of the spring fingers, differs between mechanical and hydraulic clutch control systems.

With a simple mechanical version the spring loading on the clutch pedal is reversed in direction, so that it tends to move the pedal towards the disengaged position until all free-play is taken up. The inner control cable is therefore subjected to a permanent slight tension, such that the release bearing is maintained in continuous contact with the fingers of the clutch unit. As the clutch linings wear, the fingers force the release bearing to retreat. This transmission of force, via the release fork and cable control, opposes the light spring loading on the pedal and raises it. The increase in pedal height thus signals to the driver the state of wear of the clutch linings. An additional refinement comprises a switch on the pedal mounting bracket, so that a warning light operates when the pedal rises to a certain critical height.

In the case of a hydraulic version (*Figure 2.11*), the installation and operation are similar to that of a non-continuous-contact system, except that no return-spring is fitted to the release fork. This enables the spring-loaded piston in the slave cylinder to 'float' and retreat in sympathy with the release bearing as the linings wear. It should be noted that the clutch pedal remains spring-loaded towards the engaged position, as with a non-continuous-contact system, and that a very small amount of pedal free-play is retained, typically about 1 to 5 mm. The purpose of this free-play is to ensure that the piston in the clutch control master-cylinder always fully returns to its 'off' position when the clutch is engaged, so that fluid recuperation can take place unimpeded through an uncovered compensating or by-pass port. No indication of lining wear is signalled to the driver with this type of continuous-contact clutch control, since there is neither reducing free-play nor increasing height of the clutch pedal.

In conclusion it must be said that although the clutch control systems just described are in theory 'constant contact', in practice they are 'virtually constant contact'. This is because any slight run-out of the spring fingers tends to create a minimal play at the release bearing with the clutch engaged.

2.7 CENTRIFUGALLY-OPERATED CLUTCHES

In § 2.2 of *Vehicle Technology 1* we considered the general construction of conventional plate clutches, which rely entirely on the energy stored in compressed coil and diaphragm springs to supply the necessary clamping load on their friction members. As an alternative construction, the clamping load can be derived either wholly, or partially, from energy generated by a centrifugally-operated mechanism, although clutches based on this principle are chiefly associated with past automotive practice. Such clutches may be conveniently classified under two headings, as follows:

(1) Centrifugally-assisted.
(2) Fully-centrifugal.

2.7.1 Centrifugally-assisted clutch

Originally, and still known as the 'Long' clutch in America ('Long' is a registered trademark of the Borg-Warner Corporation), this type of clutch was once widely used on American large-capacity, high-torque, engines before the advent of automatic transmission, and also found similar applications elsewhere such as on the early post-war Rolls-Royce and Bentley cars. It continues to be highly regarded by constructors of competition cars in America.

Clutches of this type represent a fairly simple modification of the conventional coil spring unit. The basis of their operation is in the use of centrifugal assistance acting in parallel with the spring loading on the pressure plate, so as to avoid incipient clutch slip during high-speed gear changes. This effect is achieved by forging off-set bob-weights integral with the outer ends of the release levers. The direction of off-set being such that, as the clutch revolves, the tendency of the bob-weights to move outwards centrifugally is reacted as an additional clamping load at the pressure-plate, via the pivoted release levers.

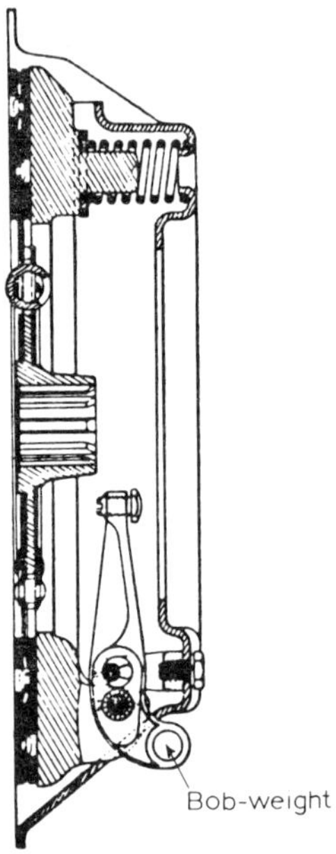

Figure 2.12 Cross-section through a centrifugally-assisted clutch

In other words, the effect is analogous to that of forcibly pulling the inner ends of the release levers away from the clutch, rather than towards it, as when the clutch is being disengaged. It is for this reason that the centrifugally-assisted clutch can be critical in respect of pedal free-play, because an amount that is just sufficient with the engine idling can be insufficient at higher speeds when the centrifugal assistance is greatest. By virtue of its centrifugal assistance, the spring-loading requirement of this type of clutch can be reduced to a level which only just exceeds that required to transmit maximum engine torque. This criterion has to be comfortably exceeded in a conventional clutch for the reason mentioned in § 2.7.2. For ordinary production cars the centrifugally-assisted clutch thus required less effort to operate during traffic driving.

Figure 2.12 shows a cross-section through a clutch of this type.

2.7.2 Fully-centrifugal clutch

In the once well-known Newton clutch of this type, the engagement of the pressure plate was effected by the action of bob-weights, each pivoting from the clutch cover and acting in series with a pair of helical control springs. The bob-weights were carried on the free ends of bell-crank levers and swung within clearance holes provided in the flywheel (*Figure 2.13*). By limiting the outward travel of the bob-weights, and hence the amount of force that could be transmitted through the control springs, it was possible to incorporate a clutch withdrawal mechanism of the form normally associated with a conventional coil-spring clutch.

If the bob-weights were allowed to exert, unimpeded, their full centrifugal effect on the engaged clutch, then it could prove well-nigh impossible to disengage the clutch with any sensible amount of effort on the clutch pedal. This is, of course, because the centrifugal loading varies as the square of rotational speed, so that the effort required to disengage the clutch at an engine speed of say 1500 rev/min would be nine times as great at 4500 rev/min. As in the case of a conventional clutch, there was no point in allowing the clamping load to exceed that required for transmitting maximum engine torque by more than about 40%. This extra margin of loading ensures that the clutch springs still exert sufficient clamping load on the pressure-plate when the linings are worn, since the compressed length of the springs then becomes greater and their load in position less.

Of more recent origin is the DAF 'roller' type of fully-centrifugal clutch (*Figure 2.14*), which at engine idle speed is held disengaged by

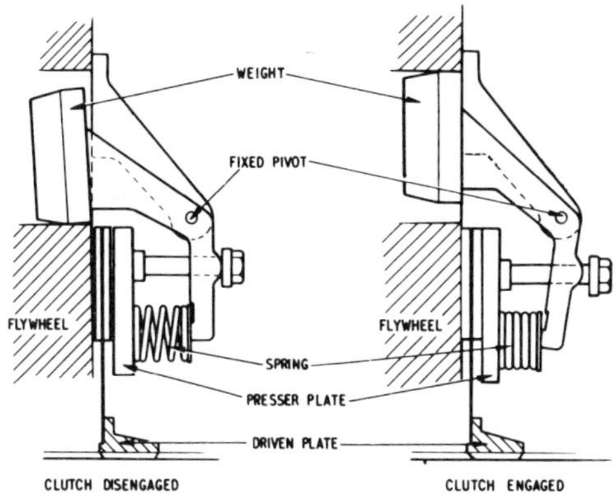

Figure 2.13 (*above*) Basic operating principle of a fully-centrifugal clutch

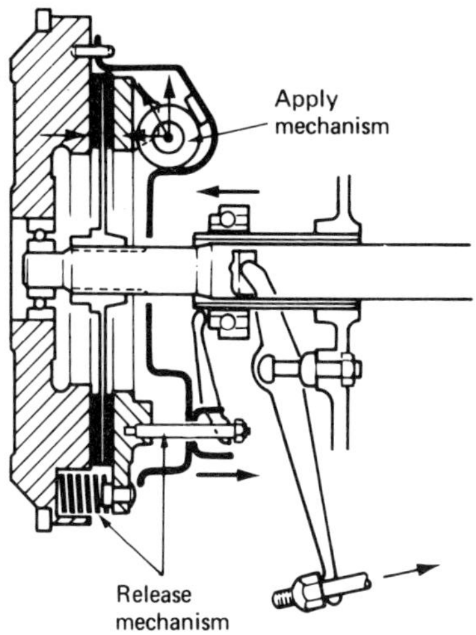

Figure 2.14 (*right*) Operating principle of roller-type of fully-centrifugal clutch

three helical springs. As engine speed is increased above idle, three rolling bob-weights are thrown outwards by centrifugal action. These bob-weights roll in converging tracks between the clutch cover and the pressure-plate and force the latter into contact with the friction-lined centre-plate, thereby causing the drive to be taken up. An advantage with this particular construction is that the engagement characteristics can be better matched to any given clutch installation, since the roller tracks can be altered in profile to modify the clamping load exerted by the bob-weight rollers.

Among the disadvantages associated with fully-centrifugal clutches in the past are the following:

(1) Clutch slip is more likely to occur if the engine is allowed to labour at low speeds.
(2) Jerky gear changes can result from the less progressive engagement characteristics of the clutch.
(3) If clutch judder arises it cannot be minimised by sympathetic driver control.

2.8 MULTI-PLATE CLUTCHES

The factors that determine the torque-transmitting capacity of a plate clutch were listed in § 2.2 of *Vehicle Technology 1*. So far we have considered the clamping force acting between the friction surfaces and the friction characteristics of the lining materials. Clearly there must be practical limitations to the extent these factors may be increased. Higher clamping forces can demand excessive driver effort to operate the clutch, while materials with higher friction values can tend to make a clutch fierce in engagement.

We are therefore left with the remaining factors of increasing the effective radius of the friction surfaces—that is, fitting a larger-diameter clutch—and increasing the number of friction surfaces in contact. Apart from any space limitation, it is not always possible simply to increase the diameter of a clutch because of the greater peripheral speed attained by the friction liners. This subjects them to higher centrifugal loading, which could tend to make them fly apart if the clutch were disengaged at high rev/min. One of the more important tests carried out on the physical properties of a clutch lining concerns its ability to withstand bursting stresses.

The final factor to be considered is therefore that of increasing the number of friction surfaces to provide a 'multi-plate' clutch. Early clutches of this type utilised a stack of driving and driven thin metal discs and were once widely used as rivals to the early cone-clutch, both types being later superseded by the single-plate type. The modern multi-plate clutch is, in contrast, usually of twin-plate construction and apart from being used on competition cars it is mainly confined in application to heavy-duty commercial vehicles.

2.8.1 A commercial vehicle twin-plate clutch

In a typical example the clutch cover assembly is produced from cast iron and is attached to the engine flywheel by a ring of set-screws. A pressure-plate of the same material is driven by means of lugs extending into mating slots of the clutch cover. Cast iron is also used as the material for the intermediate plate, which is driven by means of six 'drive pins' fitted in the engine flywheel, these being locked in position by allen screws. The pair of friction-lined driven plates are similar in construction to a conventional clutch centre-plate. Three release levers are used, their outer ends pivoting from eyebolts secured to the cover plate. Adjacent to these pivots are those that

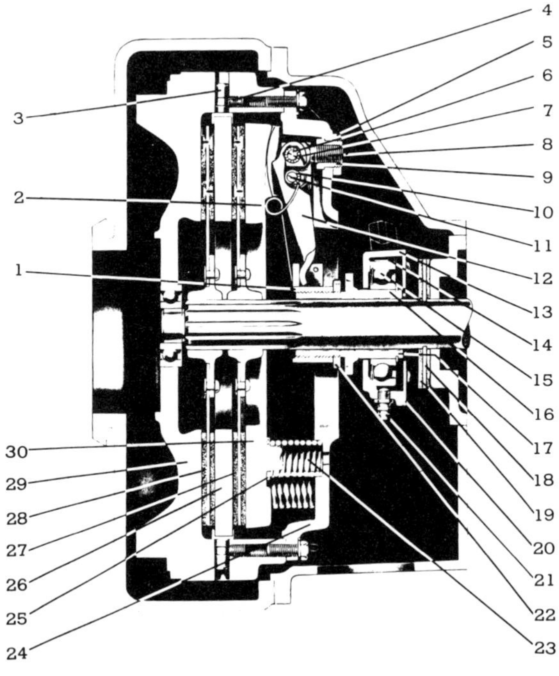

1 Release lever spider
2 Retractor spring
3 Intermediate plate drive pin
4 Socket head (Allen) set screw
5 Lock-nut
6 Eyebolt pin
7 Needle bearing (eyebolt)
8 Eyebolt
9 Adjusting nut
10 Needle bearing (pressure plate)
11 Pressure plate pin
12 Release lever
13 Release bearing housing
14 Bearing retaining spring
15 Release bearing
16 Sleeve and bushing assembly
17 Snap ring
18 Steel disc (clutch brake)
19 Friction discs (clutch brake)
20 Bearing housing cover
21 Lubrication fitting
22 Sleeve lock-nut
23 Pressure spring
24 Flywheel ring (cover)
25 Insulating washer
26 Intermediate plate
27 Driven disc assembly (PP side)
28 Driven disc assembly (FW side)
29 Engine flywheel
30 Pressure plate

Figure 2.15 Nomenclature of a heavy-duty twin-plate clutch (*Lipe-Rollway*)

connect the levers to the pressure plate. To reduce friction, needle-roller bearings are incorporated at all six pivot points of the release levers.

It should particularly be noted that the release levers operate on the 2nd, rather than the 1st system of levers, so that the clutch is disengaged by a release bearing that pulls (not pushes) on the release levers. There is a twofold advantage to be gained from this arrangement, as follows:

(1) The pedal effort is reduced for a given release load at the pressure-plate.
(2) It allows the ready incorporation of a 'clutch brake' to facilitate upward gearchanging.

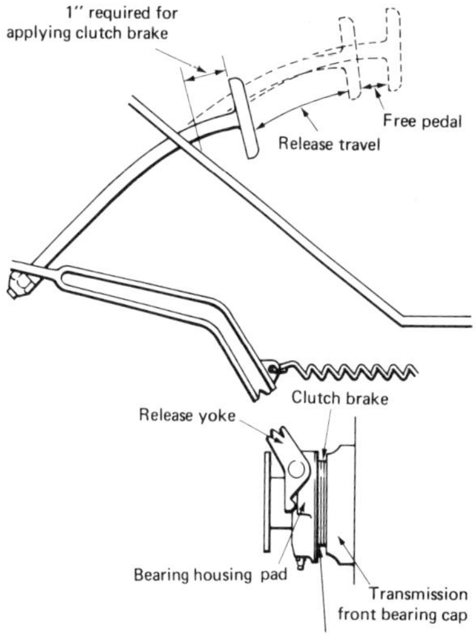

Figure 2.16 Clutch in disengaged position applying clutch brake (*Lipe-Rollway*)

The clutch brake is brought into action only when the clutch pedal is depressed to the limit of its travel, which results in the release bearing housing pad coming into rubbing contact with the brake disc on the gearbox primary shaft. Its purpose is to dissipate in the form of heat the kinetic energy of the spinning clutch driven plates, together with that of the gearbox primary and layshaft assemblies. Inevitably the twin-plate clutch possesses greater inertia than a single-plate one, so that without a clutch brake the reluctance of the driven plates to lose momentum would enforce very slow upward gear changes.

3 Manually-operated gearboxes

3.1 GEARBOX SYNCHROMESH DEVICES

When describing the constant-mesh gearbox in § 3.2.4 of *Vehicle Technology 1,* a brief reference was made to the additional refinement of a 'synchromesh' mechanism for the purpose of simplifying gear changing. Historically, this type of mechanism may be included among the many important contributions to more relaxed driving that have been pioneered by General Motors in America. The first synchromesh gearbox appeared on their 1928 model Cadillac, since when there have been many variations on the theme. In all cases the operating principle is such that gear engagement by sliding dog-clutch is preceded by frictional contact between a pair of conical surfaces which possess a high torque capacity for a relatively small size. The latter thus serve to equalise or 'synchronise' the rotational speeds of the gearwheel and its sliding dog-clutch, so that final engagement of their dog-teeth can be effected quietly. It is our purpose now to describe examples of these particularly ingenious synchromesh devices, which may be classified as follows:

(1) Conical dog.
(2) Constant load.
(3) Inertia lock.
(4) Servo ring.

3.1.1 Conical dog

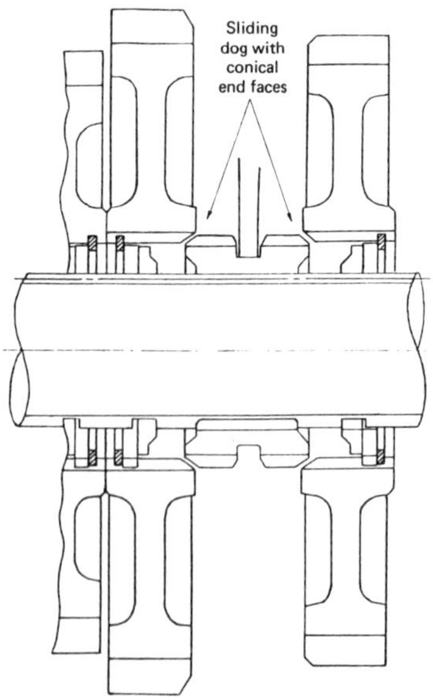

Figure 3.1 Conical-dog type of synchromesh (*Eaton*)

From a purely historical point of view, this form of synchromesh device should be last on the list, since it was introduced in the early nineteen-sixties by the Fuller Transmission Division of the Eaton Manufacturing Company for application to heavy commercial-vehicle gearboxes. The reason for describing it first is because it represents the simplest possible means of obtaining a synchronising action in the constant-mesh gearbox.

With this form of synchromesh (*Figure 3.1*) each gear to be engaged is locked to the mainshaft by a dog-clutch sliding on splines in the conventional manner, but the ends of the dog-teeth both in the gear and on the clutch are provided with matching chamfers at an angle of 35°. Before the dog-teeth finally slide into engagement, their conical end faces can therefore act in the manner of a small cone-clutch, so that any difference in relative speed between the gear and its dog-clutch may be resolved by frictional drag.

In other words, the conical dog-teeth confer a synchronising action to prevent their own premature engagement, which otherwise would result in noisy gear changing together with undesirable shock-loading on the parts concerned.

3.1.2 Constant load

It is sometimes wrongly inferred that this now obsolete type of synchromesh was the first to find application in the motor-car

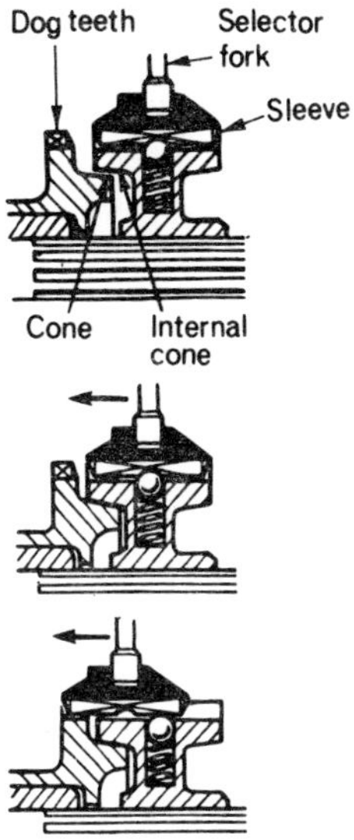

Figure 3.2 Operation of constant-load type of synchromesh

gearbox. This is possibly because its action was much simpler, but albeit less fool-proof, than that of the original 'inertia lock' synchromesh introduced by General Motors and mentioned earlier. The so-called 'constant load' form of synchromesh was, in fact, a later development by the Warner Gear Division of the American Borg-Warner Corporation, who introduced it in 1931 for their proprietary T 81 model gearbox.

This type of synchromesh (*Figure 3.2*) differed from the previously-described 'conical dog' system by physically separating the synchronising and engaging functions of the sliding dog-clutch. For this purpose the latter was made telescopic with a splined inner hub, each end of which was formed as an internal cone, and a splined outer sleeve that served also as the internal dog-teeth. A corresponding male cone and external ring of dog-teeth were carried by the gears to be engaged. The synchroniser hub and sleeve assembly were spring-loaded together by a series of radially-disposed 'detent' springs and balls that located in a groove within the sleeve.

In operation, the initial pressure exerted on the gear lever caused the selector fork to move simultaneously the synchroniser sleeve and the hub along the mainshaft, until the internal cone of the hub was brought into contact with its mating cone on the gear. Hence, the frictional drag so created synchronised the rotational speed of the gear with that of the mainshaft. Further pressure on the gear lever then overcame the restraining action of the spring-loaded balls on the outer sleeve, the resulting movement of which allowed its internal splines-cum-teeth to engage the dog-teeth on the gear wheel.

From the foregoing description it should be appreciated that the designation 'constant load' did not actually relate to the effort exerted by the driver on the gear lever, but referred to the definite load that had to be placed on the friction cones in order that sufficient synchronising action could take place. The load in question was therefore derived from the spring-loaded balls acting between the hub and sleeve. A disadvantage of the constant-load synchroniser was that this load had to be fairly high. Even for a light car it could result in a break-away force of about 200 N being required to slide the sleeve over its hub before gear engagement could be effected. Furthermore, there was nothing to stop the inexperienced driver from clashing the engaging dog-teeth by exerting too much effort on the gear lever, especially during a down-change. This difficulty was

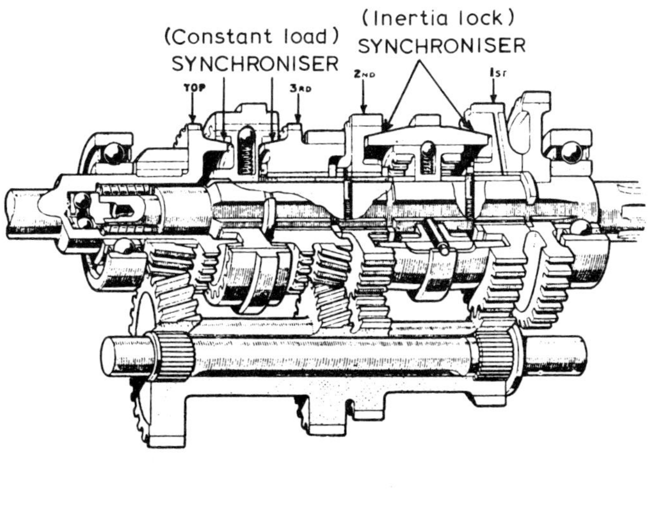

Figure 3.3 An interesting rare combination of constant-load and inertia-lock types of synchromesh once used on Hillman and Commer vehicles

recognised in one early design of all-synchromesh gearbox where only the 3rd and top gears had constant-load synchronisers, the more difficult to engage 2nd and bottom gears being provided with inertia-lock ones.

It is perhaps of interest to recall that for sports cars the constant-load synchromesh gearbox was once thought preferable to the virtually fool-proof inertia-lock type, quite possibly because their drivers were better practised in correctly timing a gear change and had no need for synchromesh assistance anyway!

Figure 3.3 illustrates a combination of constant-load and inertia-lock types of synchromesh.

3.1.3 Inertia lock

With the constant-load type of synchromesh it is possible for engagement of the dog-teeth to take place before synchronisation is complete, as already explained. This is virtually impossible with the inertia-lock type of synchromesh (*Figure 3.4*), because while friction torque exists between the mating cones of the sliding dog-clutch and the gearwheel to be engaged, the sliding dog-teeth are positively prevented or 'baulked' from engaging those on the gearwheel. For this purpose 'delay action' inclined surfaces must be incorporated in series with the friction cones and the sliding dog-clutch to provide an 'inertia lock' facility between them. The inertia lock remains active until all friction torque between the mating cones disappears, following which the sliding dog-teeth are no longer restrained from engaging those on the gear wheel and the gear change is completed.

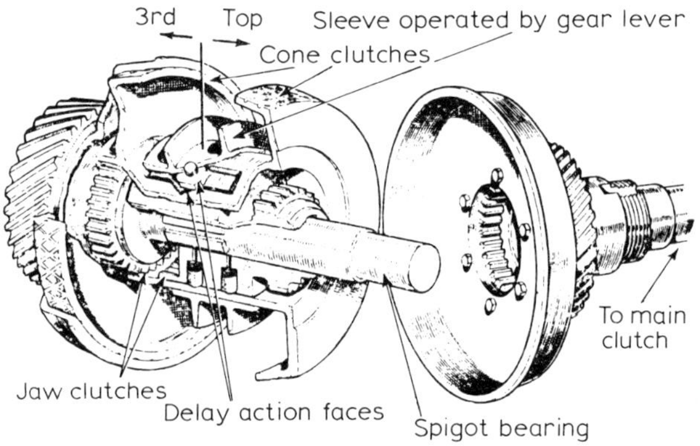

Figure 3.4 A classic example of inertia-lock synchromesh with radial blocking pins once used on Rolls-Royce cars

As indicated earlier, there are many variations on the original inertia lock theme, which may be sub-classified as follows:

(1) Blocking pin.
(2) Baulking ring.

Blocking-pin synchronisers: These represent the earliest forms of inertia lock devices used by General Motors of America and also licensed to certain other manufacturers at that time. The first example to be illustrated (*Figure 3.4*) shows the blocking pins disposed radially between the sliding dog-clutch and a slotted outer sleeve that carries the friction cones. Spring-loaded ball detents are fitted between the hub and the sleeve. During gear selection the whole assembly of hub and sleeve slides axially along the mainshaft splines until the friction cones engage. If the synchromesh hub and gear are not rotating at the same speed, the friction torque at the cones causes the slotted sleeve to turn slightly relative to the hub, until the delay-action faces of the blocker pins register in the V-shaped recesses of the slots. This effectively locks the sliding-dog hub against further endwise travel.

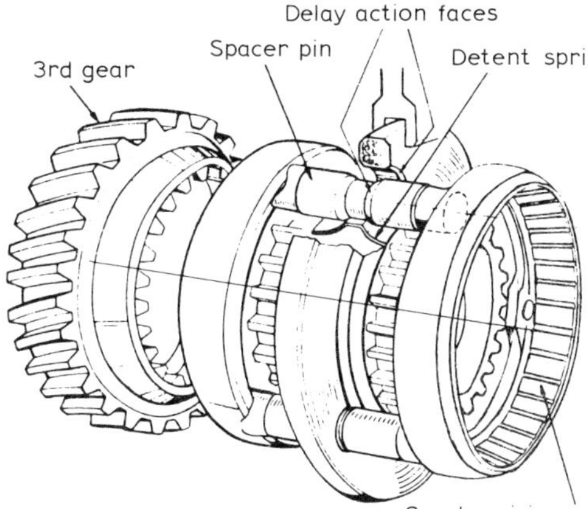

Figure 3.5 An example of inertia-lock synchromesh with axial blocking pins (*Smiths Industries*)

With continued pressure on the gear lever the hub blocker pins produce a 'centralising torque' on the slotted sleeve, so that the dog-clutch teeth may proceed into engagement with those on the gear wheel. This can occur, however, only when the centralising torque exceeds the 'synchronising torque' at the friction cones—or in other words, when all slipping between them has ceased and there is no longer any difference in rotational speeds of the synchromesh hub and the gear wheel to be engaged. At this point the blocking pins are free to ride out of their recesses in the slotted sleeve, so the spring-loaded ball detents can be overriden and the sliding dog-clutch moved into engagement to complete the gear change.

This particular version of blocking-pin synchroniser lends itself to a robust form of construction suitable for heavy-duty applications. It may also be regarded by many service engineers as representing something of a 'Chinese puzzle' to assemble! In the second example to be illustrated (*Figure 3.5*), the synchroniser friction cones receive their support from axially-disposed blocking pins spaced around the inner hub comprising the sliding dog-clutch. The latter is provided with a relatively large diameter central flanged portion, this being drilled with a series of holes such that it can slide either way over the blocking pins, as directed by the gear selector fork. A system of detent spring loading is incorporated between the assembly comprising the blocking pins and cones and the sliding dog-clutch.

As so far described such a device would simply provide a constant-load synchromesh action. Therefore in order to introduce an inertia lock effect to avoid premature engagement and clashing of the dog-teeth, the blocking pins have a central waisted portion with delay action chamfers at each end. The corresponding blocking pin clearance holes in the hub flange are similarly chamfered. During gear selection the whole assembly of flanged hub, blocking pins and cones slides axially along the mainshaft splines until the friction cones engage. If the synchromesh hub and gear wheel are not rotating at the same speed, the friction torque at the cones causes the assembly of blocking pins and cones to turn slightly relative to the hub, until the waisted portions of the blocking pins come into contact with one side of their chamfered holes in the hub flange. The sliding dog-clutch is therefore effectively locked against further

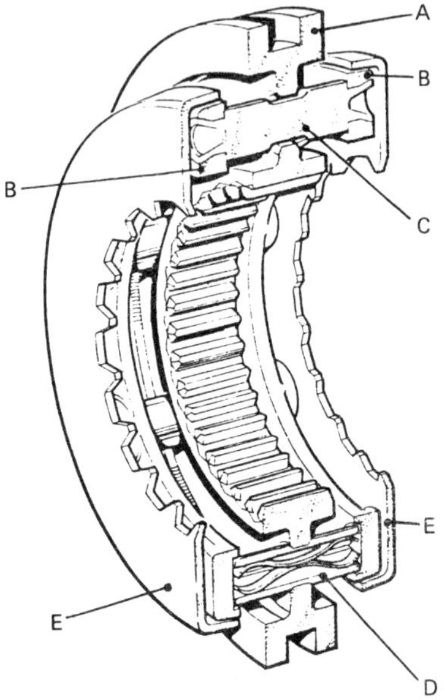

A — Synchro hub
B — Synchro rings
C — Blocker pins
D — Split energiser pins
E — Synchro cups

Figure 3.6 A modern inertia-lock synchroniser with axial blocking pins (*Turner-Spicer*)

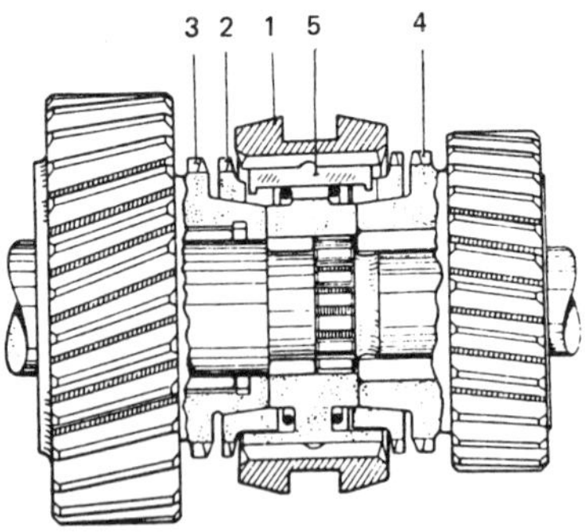

1 Synchroniser sleeve
2 Bronze baulking ring
3 Dog clutch, 3rd-speed gear
4 Dog clutch, 4th-speed gear
5 Spring-loaded shifting plate

Figure 3.7 A conventional baulking-ring type of synchromesh (*SAAB*)

endwise movement. With continued pressure on the gear lever, the chamfers on the blocking pins act against those of the hub flange holes to produce a centralising torque on the assembly of blocking pins and cones, so that the hub dog-teeth may proceed into engagement with those on the gear wheel. As in the previous case, it is not until the friction torque at the cones falls to zero that the centralising torque can exceed the synchronising torque. Then the hub flange chamfers riding on those of the blocking pins move the latter aside and allows the sliding dog-clutch to override the detent spring loading and engage the dog-teeth of the gear.

This type of blocking-pin, inertia-lock, synchroniser continues to find favour mainly for heavy-duty applications, because it permits a construction with compact overall dimensions and has the ability to operate for high mileages without attention. A notable design in this field is that produced by Turner-Spicer and known as the 'split-pin' synchroniser (*Figure 3.6*). The adjacent synchromesh cones are linked to each other by three blocking pins in the usual manner, but these are supplemented by three 'split pins' that provide the detent spring loading. Each split pin is made in two halves with a square centre section into which two specially-shaped springs are assembled back-to-back. The particular advantage of this form of construction is that it eliminates sliding friction on the detent springs themselves.

Baulking ring synchronisers: These may be said to represent the later forms of inertia-lock devices of which the classic 'baulk ring' design, introduced by the American Warner Gear Division of the Borg-Warner Corporation in 1939, remains in production worldwide. An interesting variation on this theme is the 'inverted cone' type of baulking ring synchromesh introduced by the Italian Fiat Company in 1967.

The first example of a conventional baulk-ring synchroniser to be illustrated (*Figure 3.7*) may appear at first sight to be very similar to the earlier Warner constant-load type. Closer examination will show, however, that the splined outer sleeve actuates three spring-loaded detent 'shifting plates'. These can be moved along the inner hub, which is splined and held axially on the mainshaft. The ends of the shifter plates engage with corresponding slots in the adjacent back faces of externally toothed 'baulk rings', but the width of the slots is such that the rings may turn to the extent of being one half-a-tooth out of alignment with the sleeve splines. As in the case of the constant-load synchroniser, the sleeve splines also serve as the internal teeth of the sliding dog-clutch, except that now their chamfered ends form delay-action faces with those of the teeth on the baulk rings. The latter are tapered internally to act as the synchroniser friction cones and they are generally manufactured from a copper-based alloy material such as manganese bronze.

During gear selection, the splined outer sleeve together with shifter plates is moved along the inner hub until the plates butt against the bottom of the slots in the baulk ring. This is then brought into contact with its mating cone on the gear wheel to be engaged. If the synchroniser hub assembly and gear wheel are not rotating at the same speed, the friction torque at the cones causes the baulk ring to turn slightly relative both to the hub and its outer sleeve. Since the sleeve dog teeth and those on the baulk ring are now axially misaligned, the sliding dog-clutch function of the synchroniser is temporarily halted. With continued pressure on the gear lever, the delay-action chamfers on the teeth of the hub sleeve and baulk ring produce a centralising torque that is trying to align the two sets of dog-teeth. Again, the centralising torque can only exceed the synchronising torque when all frictional drag ceases between the mating cones. When this occurs the detent spring loading on the shifter plates is overridden and the baulk ring teeth move aside and allow those on the hub sleeve to slide past and engage the dog-teeth on the gearwheel, thus completing the gear change (*Figure 3.8*).

In another fairly widely used version of conventional baulk-ring synchromesh, the back faces of the rings are provided with lug projections rather than slots, the lugs riding in extra width slots in the inner hub. End-thrust on the baulk rings is transmitted from the sliding sleeve by detent balls and springs acting via a 'follower piece' that can be moved along the hub. This particular form of construction permits a stronger (non-slotted) baulk ring at the expense of a longer synchroniser.

The third illustration of a baulk-ring synchromesh is that of the earlier-mentioned 'inverted cone' type (*Figure 3.9*). This differs by virtue of the baulk rings having internal, rather than external, teeth and being carried on splined extensions of the gear wheels. Each baulk ring is spring-loaded away from its gearwheel and retained against the ends of the splines by a locking ring. The spline ends and the teeth of the baulk ring are chamfered to provide delay-action faces and a limited rotational movement for the ring. No axial freedom is required for the inner hub of the synchroniser, which is rigidly held on the mainshaft and simply carries a splined outer sleeve with each end formed as an internal cone.

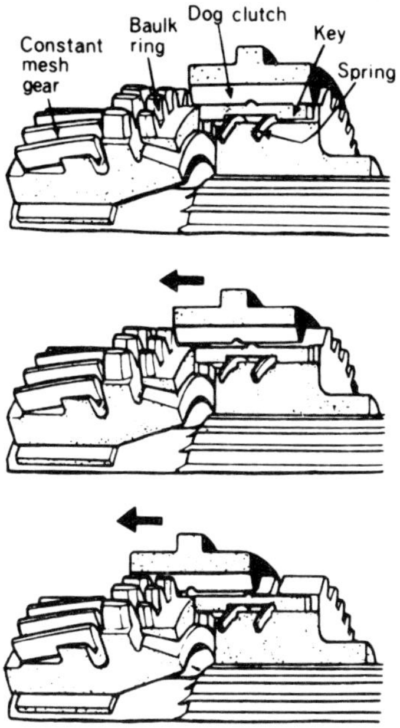

Figure 3.8 Operation of baulking-ring synchroniser

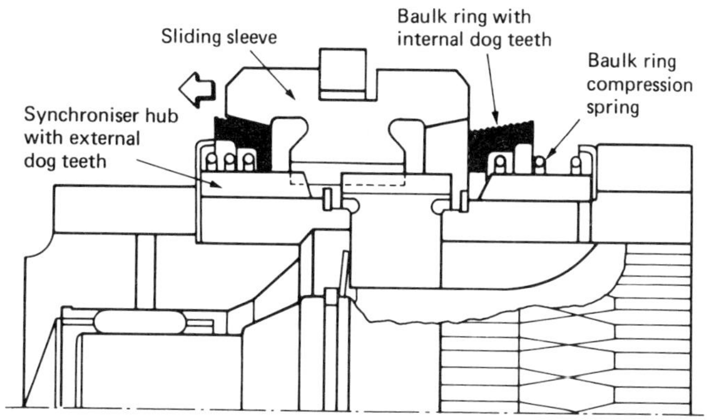

Figure 3.9 Operation of inverted-cone type of baulking-ring synchromesh (*Lada*)

During gear selection, the splined outer sleeve is moved along the inner hub until its friction cone contacts that of the baulk ring. Since the latter is spring-loaded away from the gearwheel to be engaged, any frictional drag between the cones will cause it to turn slightly and bring the delay action chamfers on the spline ends and the baulk ring teeth into contact, thereby temporarily halting the sliding movement of the hub sleeve. With continued pressure on the gear lever a situation once again presents itself where a centralising torque is acting in opposition to the synchronising torque, except that in this case it occurs between the gear-spline ends and the baulk-ring teeth. When this conflict is resolved the baulk-ring teeth move aside and the ring overcomes its spring loading to slide along the gear splines. Synchronisation having been achieved, the internal splines-cum-dog teeth of the hub sleeve follow into engagement with those on the gear extension to complete the gear change. The advantages of the inverted cone type of baulk ring, as compared to the conventional arrangement, are that its larger effective diameter of friction cone not only augments the synchronising torque, but also minimises wear and tear.

3.1.4 Servo ring

This type of synchronising device was developed by the German Porsche Company and in its original form first appeared on their 1953 Type 356 cars, which were based on Volkswagen mechanical units. Since in the first instance it was intended to occupy no more space than that required for gears without synchromesh engagement, the servo-ring synchroniser has always enjoyed the advantage of being a very compact design.

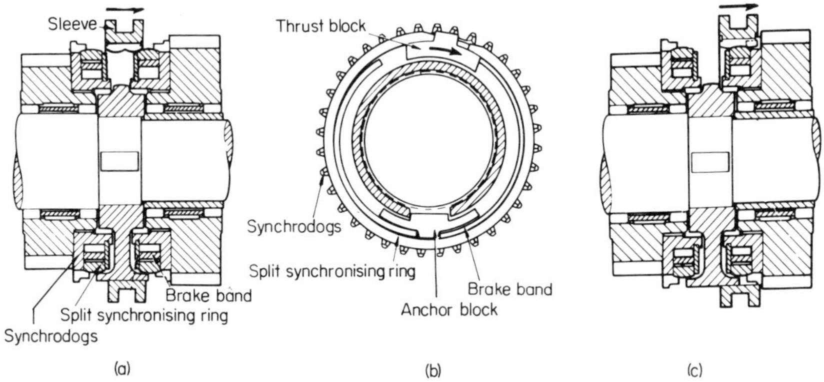

Figure 3.10 Operation of servo-ring type of synchromesh. (a)(b) Synchronising, (c) coupled

The basic operational difference between the servo ring and other types of synchroniser is that the inner cone of each mating pair assumes a 'split-ring' form, which is brought directly into contact with the chamfered splines or internal dog-teeth of the hub sleeve. if the latter is not rotating at the same speed as the split ring (driven by the gear to be engaged), then the initial frictional drag between them tends to unwrap the ring—rather like trying to push an expanded circlip round in its groove by its leading end—causing yet more friction and an increasingly powerful 'servo' or synchronising action.

Its operation was not, however, made clash-proof until the system was further developed in 1959 to include a baulking feature. The action of this later version of servo-ring synchroniser is perhaps better explained by illustration (*Figure 3.10*) rather than words.

3.1.5 Other considerations

Before leaving the topic of gear synchronising devices, it should be realised that the taken-for-granted quick, positive action gear changes associated with the typically modern baulk-ring gearbox is the result of much attention to detail by designers and including the following items:

(1) Correct choice of friction-cone angles—These are usually of 12 or 14° included angle, a smaller angle further increases the synchronising torque, but could lead to difficulty in separation of the cone surfaces.
(2) Optimum choice of delay-action chamfer angles—These usually range from 105 to 125°. included angle. A too-small chamfer angle could lead to clashing engagement of the dog-teeth, while a too-large angle could result in difficult gear changing.

(3) Provision of a suitable screw-thread finish on the conical surface of the baulk ring, this being to ensure quick dispersal of the oil film for effective synchronisation.
(4) Careful determination of detent spring loading in the synchroniser which, if too great, would result in an excessively heavy gearchange.
(5) Suitable choice of lubricants—In particular their viscosity should be low enough for them to be readily wiped off the friction cone surfaces, so as to avoid premature gear engagement and clashing of the dog-teeth.

3.2 GEARBOX HELICAL GEARS

When two gears meshing with each other are used to connect parallel shafts, and their teeth are straight and parallel to the shaft axes, we call them 'spur gears'. Such a pair of gears was illustrated in *Figure 3.2* of *Vehicle Technology 1*, wherein their application to sliding-mesh gear engagement was inferred.

If, on the other hand, the pair of gears have teeth that are inclined to the axes of the shafts, forming part of a helix, then we describe them as 'helical gears'. These are normally retained in constant-mesh engagement, the principle of which was described in § 3.2.4 of *Vehicle Technology 1.* It is possible to have a sliding mesh helical gear, but only by introducing the expensive complication of helical splines to mount it. Since constant-mesh gear engagement is used exclusively for the forward speeds of all car and many commercial-vehicle gearboxes, clearly there is a choice of using either spur or helical gears. The reasons for generally preferring helical gears are basically as follows:

(1) For the same standard of accuracy they are quieter running.
(2) For the same width they are capable of greater load carrying capacity.

These advantages are realised with helical gears so long as there is sufficient face width available to give a continuous helical action, as the load is transferred from one group of teeth to another. In other words, there is no sudden transference of load from one tooth to the next, as occurs in ordinary spur gearing. The load is gradually put on a tooth and as gradually taken off, so that the stress in any one tooth is maintained practically constant. In contrast, the stress in a straight tooth of a spur gear fluctuates according to its relative position of engagement.

3.2.1 Practical application of the helical gear

A problem that arises when single helical gears are employed is that provision must be made to absorb an end-thrust force which is imposed by the obliquity of their teeth. This 'end-thrust force' is in addition to the 'separating force' that in any event acts between two gears transmitting torque. As we shall next be dealing with the various types of rolling bearings that are used to contain these forces, it may be useful to consider in simple terms how they are generated.

Let us start with a meshing pair of straight-tooth spur gears. The tooth form of these gears is based on what is known in geometry as an 'involute curve', which was first applied to the teeth of gears by the French scientist De la Hire as long ago as 1694. It is traditional to imagine an involute as a curve that is described by the knotted end of a cord being unwound from the circumference of a circle. This circle is known as the 'base circle' (*Figure 3.11(a)*).

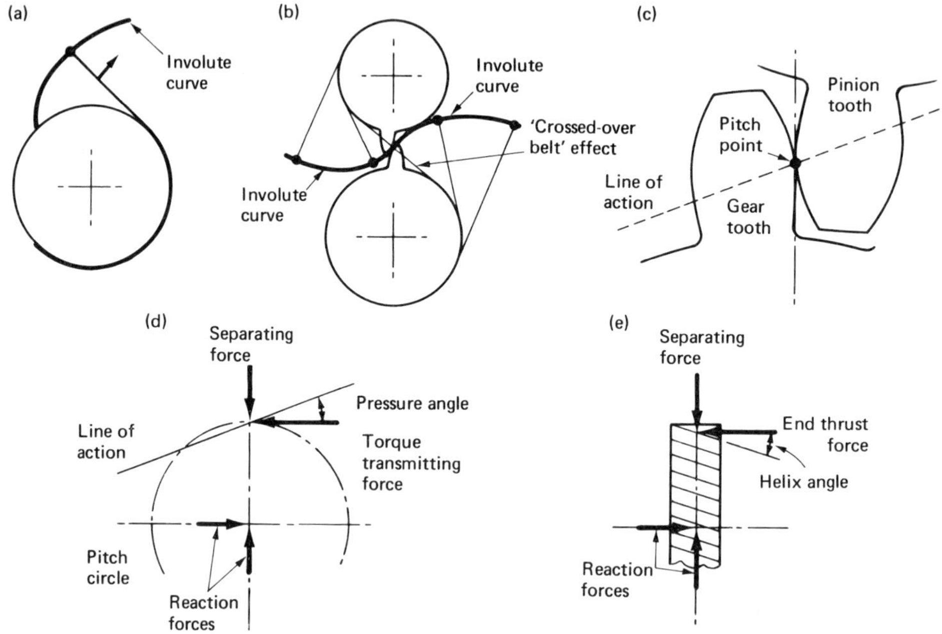

Figure 3.11 A summary of basic gear-tooth action. (a) Involute curve generated by unwinding a knotted cord, (b) action of involute gear teeth, (c) close-up of gear-teeth action, (d) radial separating force acting on a spur or a helical gear, (e) radial separating and axial thrust forces acting on a helical gear

In the case of a meshing pair of gears, their action may be visualised as being the same as if their base circles were simply connected by a crossed-over belt (*Figure 3.11(b)*). The point at which the belt crosses over being designated the 'pitch point' of the teeth (*Figure 3.11(c)*), and the straight portion of the belt not in contact with the base circles representing the 'line of action'. It thus follows that the pressure on the teeth of involute gears is constantly in the direction of the line of action and this guarantees their uniformity of motion. In other words, when two meshing gears are transmitting torque, the resultant force acting normal to the profile of each tooth occurs along the line of action (*Figure 3.11(d)*). Since the latter lies at an acute angle, termed the 'pressure angle', to the common tangent to the pitch circles of the gears, it can be resolved into two mutually-perpendicular components. Namely, a tangential force that is responsible for transmitting torque and a radial force that is trying to separate the gears (*Figure 3.11(e)*). We have therefore accounted for the 'separating force' that occurs between a pair of straight-tooth spur gears and similarly, of course, between a pair of helical gears.

It now remains to account for the additional end-thrust force generated by the latter type of gear. As mentioned previously, this end-thrust arises from the obliquity of their teeth, which is termed the 'helix angle' and is measured from the axis of the gear. The force acting along the helical teeth of the gears can therefore be resolved into two mutually perpendicular components, these being the earlier-mentioned tangential force that is responsible for transmitting torque and an axial force that is producing an 'end-thrust'. This can act in either direction and depends upon the hand of the helix (right or left), the direction of rotation and whether the particular gear is driving or being driven.

3.3 GEARBOX ROLLING BEARINGS

The selection of suitable types of bearing for the motor vehicle gearbox is based on considerations similar to those for any other engineering application. Each type of bearing has characteristic features which make it particularly suitable for certain duties, so its selection must take into account at least the following basic factors:

(1) Available space.
(2) Magnitude of load.
(3) Direction of load.
(4) Axial displacement.

Available space: In many instances at least one of the main dimensions of the bearing (usually the bore) is predetermined by the design of the gearbox shafts. If radial space is particularly limited, such as in the case of the spigot bearing in the constant-mesh pinion of the primary shaft, then a needle-roller bearing with a small sectional height may be required.
Magnitude of load: This is normally the most important factor in determining the size of bearing required, together with the related factors of life and reliability. Generally, roller bearings can carry greater loads than ball bearings of the same external dimensions.
Direction of load: Cylindrical roller bearings having one ring without flanges and needle-roller bearings can carry only radial loads. All other types of radial bearing can carry a radial and an axial load acting simultaneously or what is termed a 'combined load'. The most important feature affecting the ability of a bearing to carry an axial load is its angle of contact, which was referred to earlier in connection with clutch release ball bearings in § 2.4.2.
Axial displacement: The normal bearing arrangement consists of a locating (fixed) bearing and a non-locating (free) bearing, so that the latter can be displaced axially. This prevents cross-location of the bearings as a result of shaft expansion or contraction. The bearings can also be arranged so that axial location is provided by each bearing in one direction only, as for example with the tapered roller type, which is further discussed in connection with final drives in § 5.1.2.

3.3.1 Practical application of gearbox bearings

In a conventional double-stage, constant-mesh gearbox, as shown in *Figure 3.7* of *Vehicle Technology 1*, there is virtually no load on the gearbox bearings when direct drive top gear is engaged. It is possible that some load may be imposed on the mainshaft rear bearing, but this arises from propeller-shaft operation.

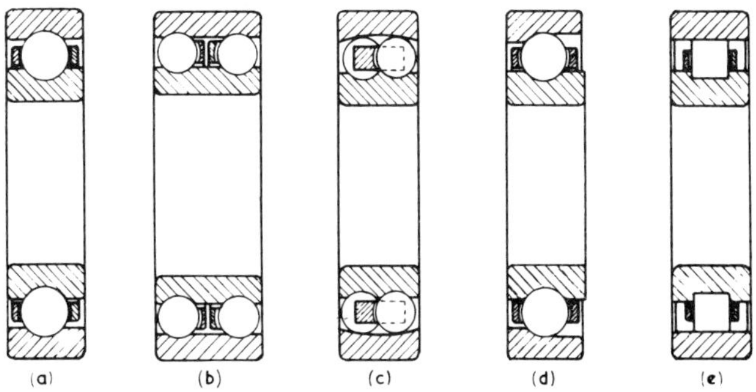

Figure 3.12 Simple ball-and-roller bearings. (a) Standard ball journal, (b) double-row standard ball bearing, (c) double-row, self-aligning ball bearings, (d) combined radial and thrust (one direction), (e) standard roller journal

Figure 3.13 Ball and needle-roller bearings in a modern five-speed commercial vehicle gearbox (*Turner-Spicer*)

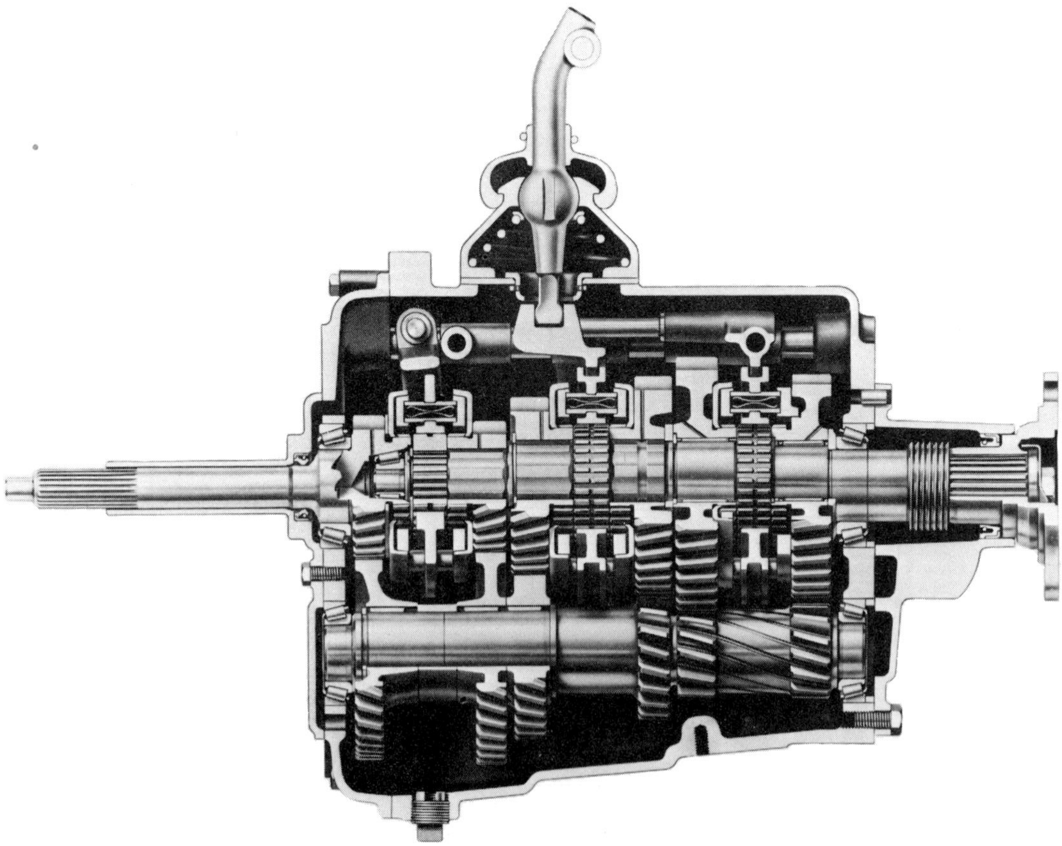

Figure 3.14 Tapered-roller bearings in a modern five-speed commercial vehicle gearbox (*Turner-Spicer*)

When the indirect gears are selected, it is then that the separating and end-thrust forces acting on the helical constant-mesh, and whichever other pair of gears happens to be engaged, have to be taken into account. This means that there will be combined radial and axial loading on both the primary shaft bearing and the mainshaft rear bearing, together with a radial load on the mainshaft front or spigot bearing. Although the layshaft bearings will be subjected to radial loading, they can be relieved of axial loading (or nearly so) by providing the driven constant-mesh gear and the driving pinions with helices of the same hand, so the two end-thrusts on the layshaft are in opposite directions. A careful choice of helix angles for the layshaft gear and pinions also plays a part in achieving this balance of loading.

Until recently, it was long established practice to use single-row, deep-groove, ball-bearing races on the primary shaft and mainshaft, in conjunction with a needle-roller spigot bearing. Needle-roller bearings (*Figure 3.13*) were also typically used to support a hollow layshaft with plain thrust washers to take any end-thrust loading. In current practice, there is a greater tendency to employ tapered-roller bearings throughout (*Figure 3.14*), the reasons for this including those of more compact design and simpler assembly. The closer control over endwise movements of the primary shaft and mainshaft is also claimed to improve the quality of gearchange.

3.4 DIRECT TYPE OF GEAR LEVER CONTROL

This relatively early type of gear level control was provided with a ball-and-socket mounting, directly in the tower of the gearbox top

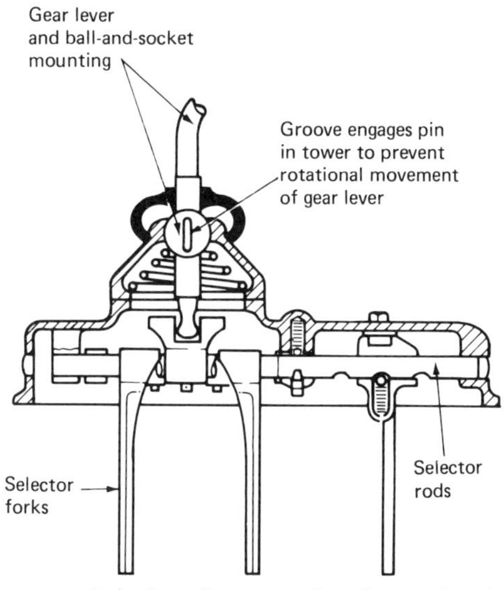

Figure 3.15 Direct type of gear lever control for commercial vehicle

cover. It is therefore associated entirely with multi-rail selection of the gears as described in § 3.3.1 of *Vehicle Technology 1. Figure 3.11* therein also showed the arrangement in principle, so it only remains to illustrate an example from actual practice (*Figure 3.15*).

3.5 REMOTE TYPE OF GEAR LEVER CONTROL

This later type of gear lever control (*Figure 3.16*) is required where the gearbox is mounted so far forward, or rearward, of the seating arrangements for the driver that it is no longer possible to use a

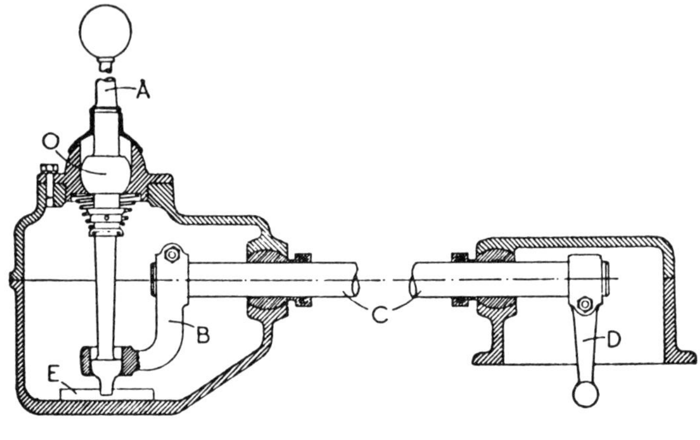

Figure 3.16 Operating principle of remote type of gear lever. A — gear lever, B — cranked arm, C — remote-control rod, D — selector finger, E — lift lever reverse guard, O — ball-and-socket joint

direct-type of gear lever. It comprises basically a remote 'push-pull' rod that is usually cranked downwards where it engages the lower end of the gear lever. Alternatively, a toggle connection may be used between the remote-control rod and the gear lever. In some installations it is necessary for the control rod to have universally-jointed ends. These various features enable the rod to transmit both axial and rotational movements to the gear selector mechanism, in sympathy with fore and aft and sideways rocking movements of the gear lever. At its end remote from the gear lever, the control rod actuates either a multi- or a single-rail gear selection system, as earlier described in §§ 3.3.1 and 3.3.2 of *Vehicle Technology 1.* In the case of a transversely-mounted engine and gearbox (*Figure 3.17*), it typically

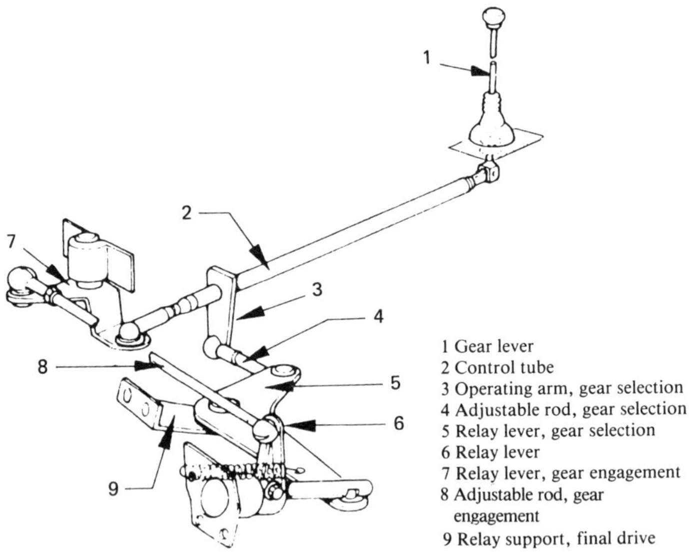

Figure 3.17 Remote type of gear-lever control for transversely-mounted gearbox (*Talbot*)

becomes necessary to introduce two bell-crank 'relay' levers, so that the control rod movements can be turned through a right-angle. One of the levers serves to engage the selector finger with the appropriate selector rod and the other moves the selector finger and the selector rod to engage the gear.

3.6 UTILISATION OF REMOTE GEAR LEVER CONTROL

A remote gear lever control is usually associated with the following transmission layouts:

(1) Forward-mounted engine and gearbox, rear-wheel drive.
(2) Front engine with rear transaxle.
(3) Front engine, front-wheel drive.
(4) Rear engine, rear-wheel drive.
(5) Front engine with forward-mounted cab (commercial vehicle).

The remote gear lever control does not necessarily have to be floor-mounted, since less commonly it may be operated from the facia or the steering column. For example, the former type of control makes for a less complicated linkage in those versions of front-wheel drive where the transaxle, combining the gearbox and final drive, is mounted ahead of the engine. It comprises simply a 'walking stick' twist to push-and-pull control rod, which is cranked where it intercepts the top of a direct-mounted gear lever.

The steering column type of gear change control (*Figure 3.18*) was an American development of the late nineteen-thirties, following its

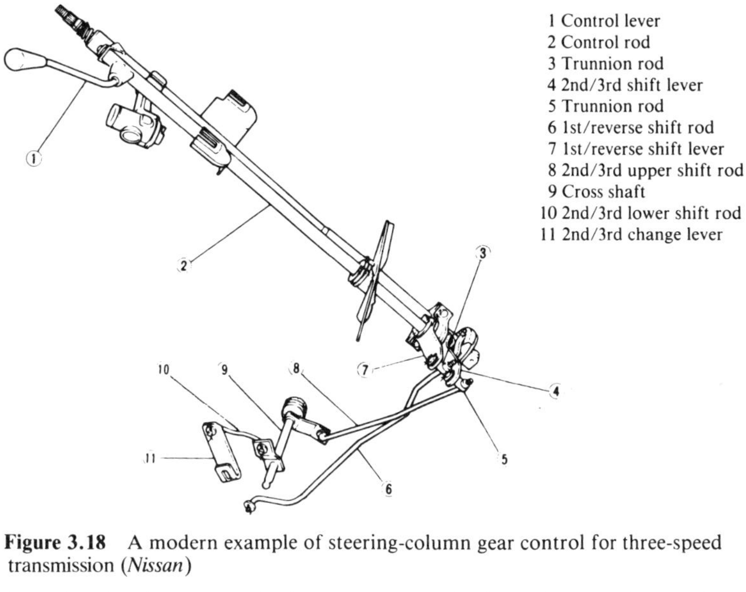

Figure 3.18 A modern example of steering-column gear control for three-speed transmission (*Nissan*)

introduction by General Motors on their 1938-model Pontiac, although the basic idea was pursued by designers of some very early cars. Its purpose was to give the front seat passengers more space and leg room and also make possible a three-abreast bench seating arrangement.

For these reasons the steering column gearchange enjoyed a similar vogue of popularity in this country during the late nineteen-forties through to the early nineteen-sixties. However, by present-day standards of gear control many such installations lacked precision of operation (to put it kindly!) because of the complexity of their linkage and the multiplicity of pivot points that were subject to wear. Also, they are not entirely compatible with the modern safety requirement for a collapsible steering column.

3.7 GEARBOX SPEEDOMETER DRIVE

The conventional magnetic speedometer is driven by a flexible drive shaft, or 'speedo cable' (*Figure 3.19*), from reduction gearing within

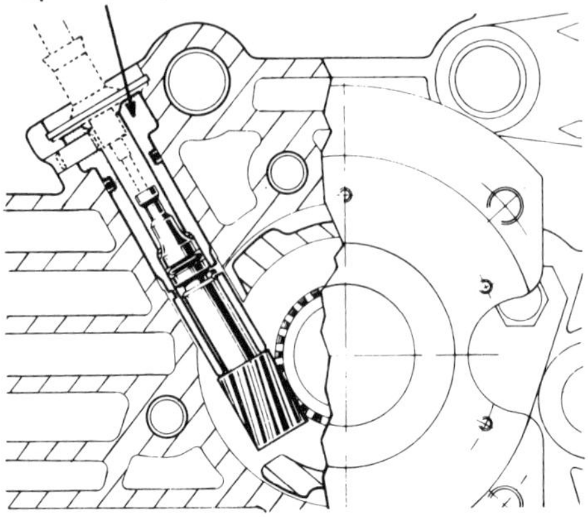

Figure 3.19 Cross-section through speedometer drive (*Alfa Romeo*)

the vehicle transmission system, such that its driving speed is always proportional to vehicle speed. With the modern car there are generally three sources from which the speedo cable may receive its right-angle drive, depending upon the layout of the transmission system, as follows:

(1) *Double-stage gearbox:* From a spiral driving gear that is either machined integrally with, or separately mounted on, a rearward extension of the mainshaft.
(2) *Single-stage (transaxle) gearbox:* In this case the spiral driving gear is clamped over a short rearward extension of the layshaft.
(3) *Transverse gearbox and final drive:* Here the spiral driving gear is mounted on the opposite end of the differential cage to the final drive helical gear.

It has long been customary to use a nylon material for the driven gear to which the speedo cable is directly connected and, more recently, also for the driving gear where this is separately mounted.

3.8 THE PURPOSE OF AN OVERDRIVE

Perhaps it may be recalled from § 3.2.6 of *Vehicle Technology 1* that with a conventional double-stage gearbox, the engine rotates at the same speed as the propeller shaft in direct top gear, but at a faster speed than the road wheels in accordance with the rear axle ratio. Now the latter necessarily represents a compromise between the following factors:

(1) It should be 'low' (numerically high) enough to give good flexibility in top gear and thus avoid the need for frequent gear changing.
(2) It should be 'high' (numerically low) enough to prevent over-revving of the engine at the maximum speed of which the car is capable in top gear. That is, the speed at which the aerodynamic and mechanical drag forces acting on the car balance the tractive effort available at the road wheels.

It is this compromise in axle ratio that can result in the engine speed being unnecessarily high when cruising at motorway speeds. The basic purpose of an overdrive is therefore to provide a system of 'step-up' gearing, which enables the engine to rotate at a slower speed than the propeller shaft in top gear. The choice of ratio for the step-up gearing is generally chosen such that the maximum speed of the car is not reduced when the overdrive ratio is engaged. Utilised to maximum advantage an overdrive can therefore permit any given journey to be completed with the least total number of engine revolutions. This confers the following benefits on vehicle operation, as will be known by all experienced drivers:

(1) Reduced fuel consumption.
(2) Quieter cruising speed.
(3) Longer engine life.

At this stage we are not required to study the actual operating mechanism of an overdrive. Suffice it to say here that it may comprise either a separately-controlled auxiliary gear unit adapted to fit on an existing gearbox, or the gearbox contains an extra pair of gears that are engaged in the normal manner. The former arrangement was essentially an American development of the late nineteen-thirties, a notable example being the Warner overdrive unit manufactured by the Borg-Warner Corporation. In England an overdrive unit originally known as the Laycock de Normanville (based on the inventions of Capt Edgar J de Normanville), became very propular in the nineteen-fifties and has continued to find application on cars and light vans. GKN Laycock are currently the world's only manufacturer of this form of unit.

In more recent years, the earlier European practice of incorporating an overdrive ratio in a conventional gearbox has undergone a general revival, especially in the form of an overdrive fifth gear with direct drive in fourth.

3.9 THE PURPOSE OF A TWO-SPEED TRANSFER GEARBOX

During World War II the Ford Motor Company and Willys Overland Motors built for the United States government a vehicle designated '1/4-ton 4 × 4', which soon became universally known as the 'Jeep' (a word derived from 'general purposes'). A distinguishing feature of this versatile military vehicle was that either the rear wheels only, or for better traction all four wheels, could be driven. It was therefore suitable both for open-road and cross-country operation

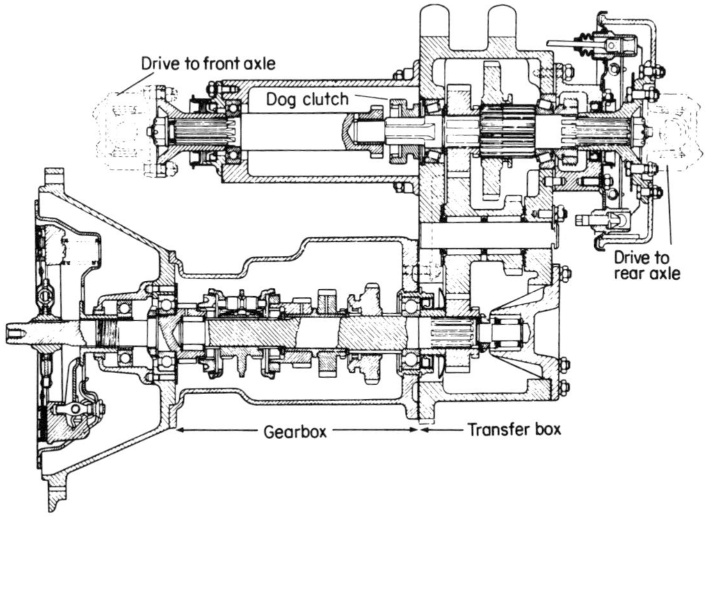

Figure 3.20 Cross-section through a two-speed transfer box (*Land-Rover*)

and it is now a matter of automotive history that it set the pattern for many other four-wheel drive (usually abbreviated to 4WD) vehicles beginning, of course, with our own equally famous Land-Rover.

Since the basic layout of the historic Jeep was similar to that of any other vehicle with a front engine driving a live rear axle, it meant that the propeller shaft drive to the live front axle had to lie alongside the engine, and this feature introduced the requirement for a 'transfer gearbox'. In fact the transfer gearbox also served another purpose, because it housed an additional pair of gears that could provide an emergency low gear ('underdrive'!) when four-wheel drive was engaged. So we therefore arrive at the designation 'two-speed transfer box', which generally implies the following duties:

(1) Provides a means of offsetting the drive to the front wheels.
(2) Allows the drive to the front wheels to be engaged or disengaged.
(3) Offers an emergency low gear with four-wheel-drive operation.

3.10 LAYOUT OF A BASIC TWO-SPEED TRANSFER BOX

A two-speed transfer box is either directly attached to the rear of the main gearbox, or is separately mounted from it and driven via a short coupling shaft. The object of the latter arrangement is to provide a centrally-disposed transfer box, so that an unduly short front propeller shaft operating with large angular movements of its universal joints is avoided.

In the modern two-speed transfer box all the gears have helical teeth and are retained in constant mesh, their engagement or disengagement being effected through a dog-clutch. For earlier designs of transfer box the low gear was usually of the sliding-mesh type with spur teeth. Another dog-clutch is provided to engage or disengage the drive to the front wheels. The chief components are as follows:

Transfer input shaft and gear: The shaft is a splined extension of the gearbox mainshaft and carries the transfer input gear. It may receive additional support from an outrigger bearing in the rear wall of the transfer box.

Intermediate shaft and gears: Similar to the layshaft cluster of some gearboxes, it comprises a fixed spindle with needle-roller bearings upon which rotate an integral idler gear and pinion.

Transfer output shaft and gears: This assembly comprises a main and an extension shaft in series with a spigot bearing between them. The main shaft is typically supported in tapered-roller bearings, which are located either side of a pair of 'high' and 'low' speed gears that are normally free to revolve upon the mainshaft. For engagement of these gears the central portion of the mainshaft is splined to carry a sliding dog-clutch. Each end of the mainshaft is also splined, the rear end to accept the flange coupling for the rear propeller shaft and the front end to carry a sliding dog-clutch for engaging the extension shaft. This in turn is splined at its front end to accept the flange coupling for the front propeller shaft.

Transfer selector mechanism: A pair of conventional rods and forks are used to perform the separate functions of engaging or disengaging the drive to the front wheels and selecting either 'high' or 'low' gear range. An interlock between the selector rods ensures that the emergency low gear can only be used in conjunction with four-wheel drive, otherwise the greater torque could overstress the rear axle's parts. The driver is provided with two extra shift levers to control the transfer box and it should be appreciated that shifting from high gear to low gear should not be attempted except when the vehicle is either being operated at very low speeds or is at a standstill.

3.11 THE OPERATION OF A TWO-SPEED TRANSFER BOX

Since this topic better lends itself to illustration, rather than description, the various modes of power-flow through a two-speed transfer box have been presented accordingly in *Figure 3.21*, and should be further related to the arrangement of propeller shafts described and illustrated in § 4.8.

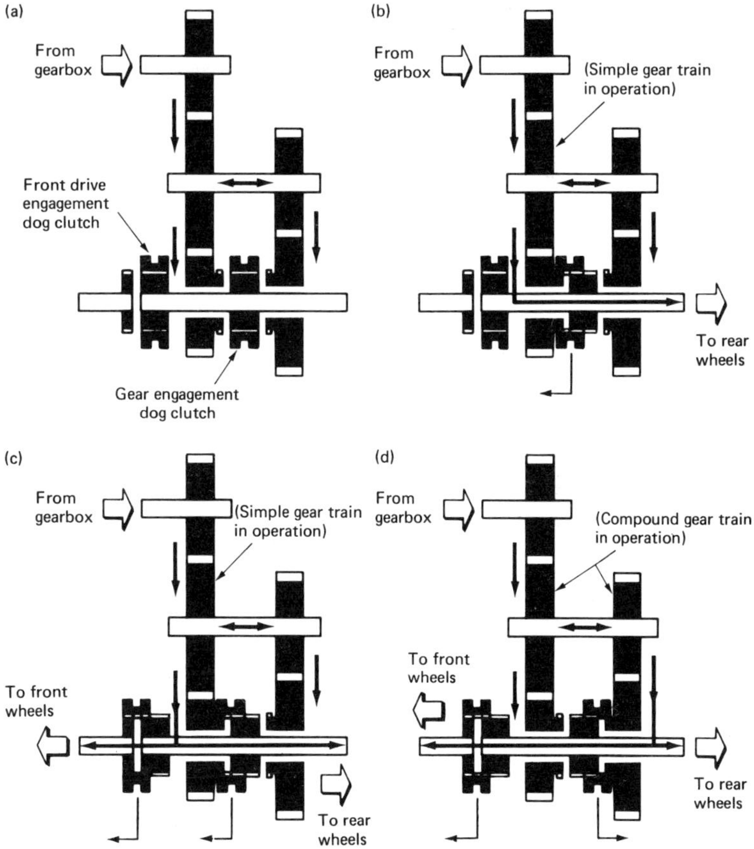

Figure 3.21 Operating principle of the two-speed transfer box. (a) No drive to wheels, (b) drive to rear wheels, (c) 'high' drive to all wheels, (d) 'low' drive to all wheels

4 Propeller and drive shafts

4.1 CONSTANT-VELOCITY UNIVERSAL JOINTS

It may be recalled from § 4.5 of *Vehicle Technology 1* that a simple Hooke's coupling, or universal joint of cross-pin construction, causes a non-uniform rotational speed to be transmitted from the driving to driven shafts when they are running with angular misalignment. It was further explained that these fluctuations in rotational speed could be cancelled by adding a second Hooke's coupling in series with the first. Although our thoughts were then concerned with a conventional propeller shaft, incorporating a simple Hooke's coupling at each end, it does of course follow that if we place two Hooke's couplings back-to-back between any pair of driving and driven shafts, running with angular misalignment, then these shafts would immediately be free of relative fluctuations in rotational speed.

In other words, the use of two back-to-back Hooke's couplings can be said to create a 'constant-velocity' universal joint of which there are also several other more modern forms. Another description of a constant-velocity universal joint is 'homokinetic' joint (literally of 'constant-velocity'), a term that was coined in the late nineteen-twenties by Automobiles Tracta who were pioneer manufacturers of front-wheel drive cars in France.

4.1.1 Transmission applications of constant-velocity joints

In modern automotive practice these can be summarised as follows:

Front-wheel-drive shafts: Certainly the earliest and perhaps the most obvious requirement for a constant-velocity joint is at the outer end of each drive shaft on front-wheel-drive cars (*Figure 4.1*). The presence of a single Hooke's joint at this point, where large angle steering movements have to be accommodated, makes for totally unacceptable 'kick-back' reactions at the steering when cornering.
Rear-wheel-drive shafts: A later and perhaps not so obvious requirement for a constant-velocity joint is at both ends of each drive shaft to independently-sprung rear wheels. Although the use of single

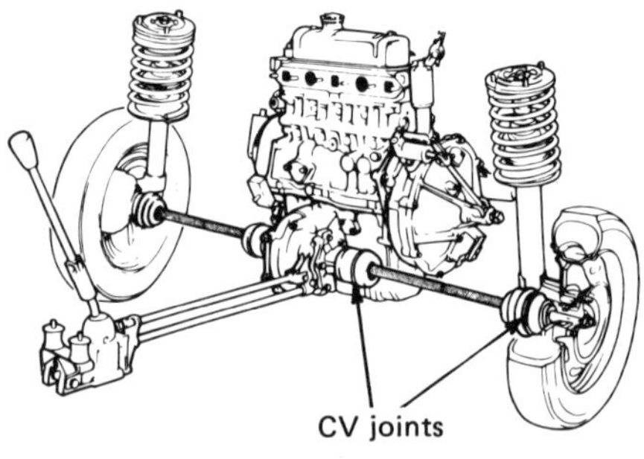

Figure 4.1 (*left*) Application of constant-velocity universal joints to FWD (*GKN Hardy Spicer*)

Figure 4.2 (*below*) Application of constant-velocity universal joints to propeller shaft (*GKN Hardy Spicer*)

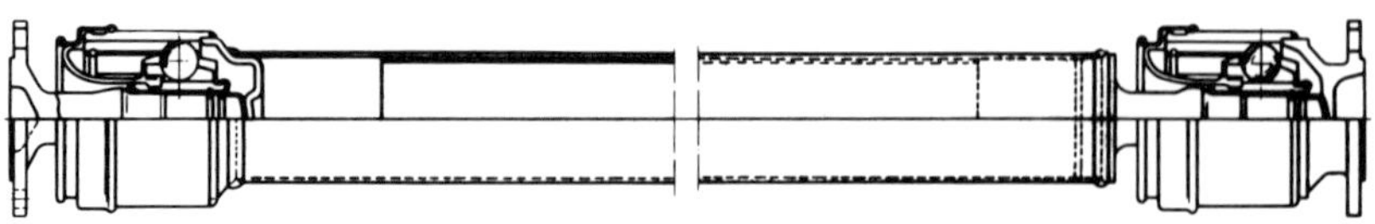

Hooke's couplings at both ends of each drive shaft does to all intents and purposes give a constant-velocity drive to the road wheels, as they rise and fall with suspension movements, there still remain the fluctuations in rotational speed of the drive shaft itself. In a modern car, the inertia effects that arise from this unevenness of drive-shaft rotation can be sufficient to produce an unacceptable level of vibration. Hence the requirement for constant-velocity joints to provide an overall uniformity of rotational movement within each drive-shaft assembly.

Rear-axle propeller shafts: Much the same considerations apply here as for rear-wheel-drive shafts. This trend began in the early nineteen-sixties in America when Cadillac fitted back-to-back Hooke's couplings to both ends of the rear part of their divided propeller shaft. The advantages claimed for this arrangement were smoother operation of the drive-line with less sensitivity to any misalignment. More modern types of constant-velocity joint are used in similar current applications and additionally perform the function of a sliding joint. This latter feature is also relevant to the first two applications mentioned above.

4.2 TYPES OF CONSTANT-VELOCITY JOINT

These may be conveniently classified under the following headings:

(1) Double-Hooke's.
(2) Tracta.
(3) Weiss.
(4) Rzeppa.

4.2.1 Double-Hooke's joint

Perhaps the best-known version of this particular form of constant-velocity joint was that introduced in 1934 by the French firm of Glaenzer-Spicer for the then new front-wheel-drive model Citroën and used for many years thereafter. An important constructional feature of this type of joint is the ball-and-socket centring device required between the driving and driven shafts (*Figure 4.3*). The basic

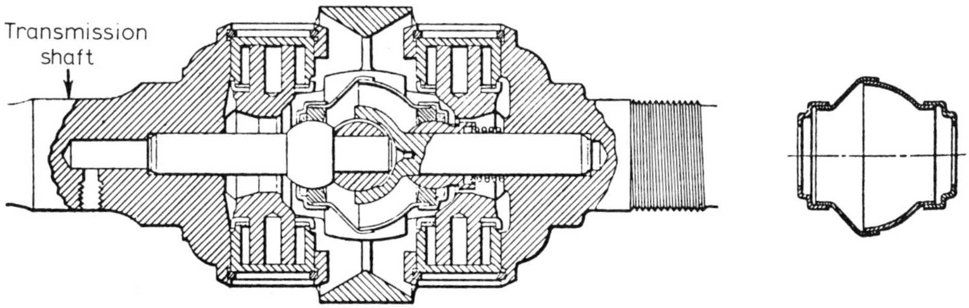

Figure 4.3 Cross-section through a double-Hooke's joint (*Citroën*)

functions of this device are those of maintaining the double cross-pin joint both self-supporting and self-aligning, and also dividing equally (or nearly so) the total joint angle between the driving and driven shafts. In performing these functions the centring device is subject to appreciable relative motion between its socket, ball and stud. Suitable arrangements for lubricating these critical parts of the double-Hooke's joint must therefore be made, typically by a grease nipple and communicating channel in the coupling yoke.

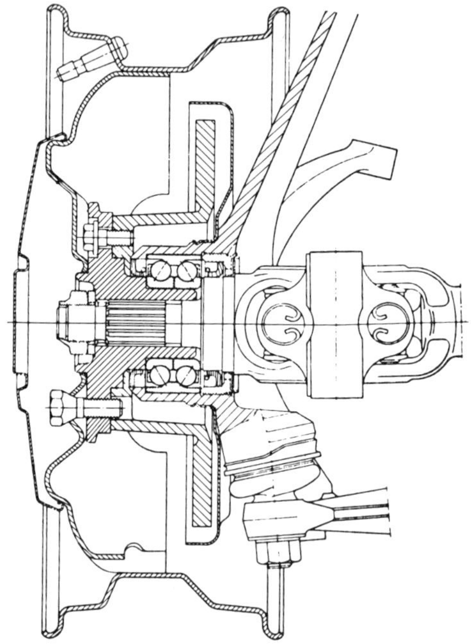

Figure 4.4 Double-Hooke's joint and hub bearing arrangement for FWD (*SKF*)

Given satisfactory behaviour of the centring device, the advantages of the double-Hooke's joint are generally those of rugged construction, durability and the ability to operate at high joint angles. Its main disadvantages are those of complicated and cumbersome construction, so that it can now be considered obsolescent for front-wheel-drive applications.

4.2.2 Tracta joint

This type of constant-velocity joint dates back to 1926 and is attributed to the Frenchman P Fenaille who, in conjunction with another pioneer designer of front-wheel-drive cars, J A Gregoire, incorporated such joints in the (previously-mentioned) Automobiles Tracta cars, hence the name of the joint. (*Figure 4.5*).

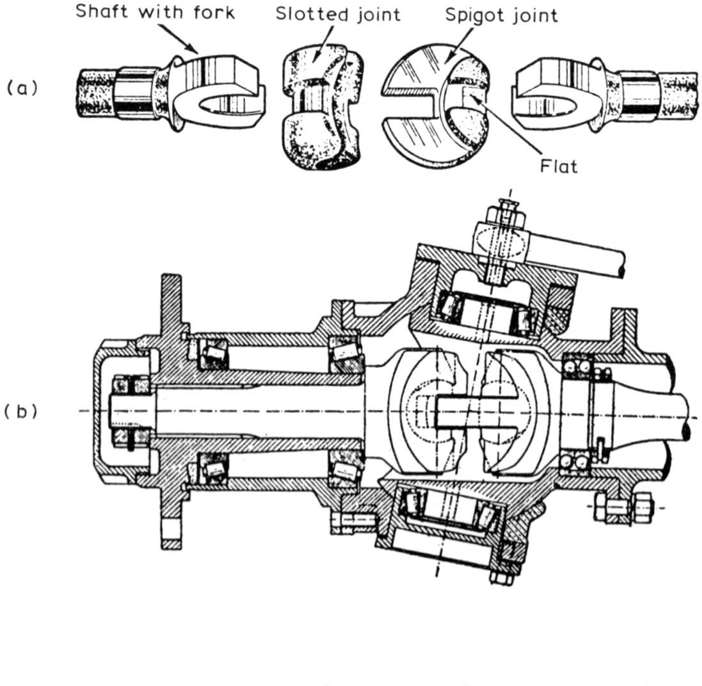

Figure 4.5 Component parts (a), and practical application (b), of the Tracta-type joint

In construction the driving and driven shafts are provided with forked ends of semicircular form, which in turn are free to articulate about a correspondingly grooved intermediate member. The latter is actually a two-piece assembly with a spigot and slotted joint that permits relative circular motion in a plane at right-angles to the forked ends of the shafts. It is thus possible for the two-piece intermediate member to adjust itself continuously to their angular displacement, thereby providing a universal action with constant-velocity drive. It will be evident that a distinguishing feature of this particular type of joint is that all relative motions occur as sliding, rather than rolling, contact between the working parts. However, the bearing surfaces are of fairly generous area and should be expected to provide a low rate of wear. On the other hand, the sliding friction at the bearing surfaces can result in overheating of the joint when articulating under high torque loading conditions and this can impose a limit on its usefulness.

Another disadvantage of this type of joint, at least for the independently-sprung and driving wheels of the modern car, is that it is not self-supporting. That is, the driven shaft requires the external bearing supports that can only really be furnished by a live-axle with steerable wheels. For this reason its later development in the automotive field was mainly confined to the driven front wheels of specialised heavy-duty and military vehicles, where it provided a simple, rugged and compact joint capable of operating at high joint angles.

4.2.3 Weiss joint

So-named after its inventor C W Weiss, who was a pioneer figure in the American diesel engine industry, it was patented by him in 1923 and represented the first of the 'ball' type constant-velocity joints. Rather interestingly the Weiss joint (*Figure 4.6*) was originally

(a)

(b)

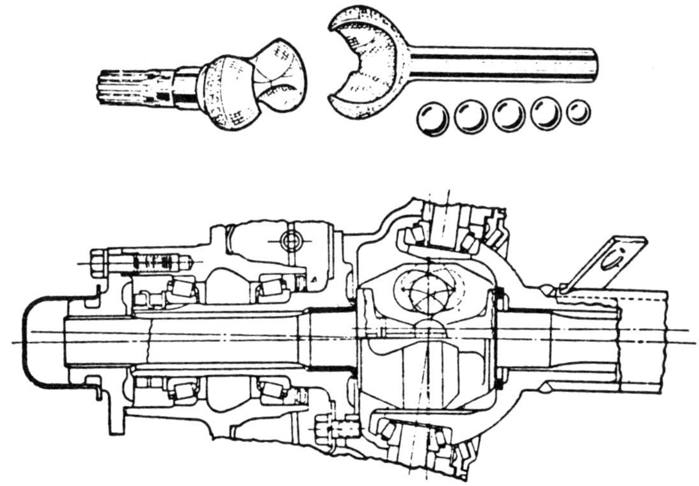

Figure 4.6 Component parts (a), and practical application (b), of the Weiss-type joint

intended to replace the enclosed single-Hooke's joint used behind the gearbox of the then popular 'torque-tube'-drive rear axles. These will be referred to again in §§ 4.6 and 4.7 of this chapter.

The innovative feature of the Weiss joint was the use of four rolling steel balls interposed between the interlocking forked ends of the driving and driven shafts. Location of the balls is by means of curved tracks machined in the opposed faces of the forked ends, each of which resembles a pair of jaws. These are separated endwise by a centring ball about which the two shafts articulate. At this juncture it is important to appreciate that the geometric centre of the curved tracks in each pair of jaws is deliberately offset from the true centre of the joint. Hence, the opposing ball tracks cross each other even when the driving and driven shafts are in alignment. The practical significance of this arrangement is that each ball has no choice but to remain at the crossing point of the curved tracks that embrace it, so that in combination the four balls must always lie in a plane which bisects the angle between the shafts, thereby providing a constant-velocity drive.

A further point to note with this type of ball constant-velocity joint is that only one-half the number of balls will transmit drive in either direction of rotation. This can result in a larger size of joint being required for any given torque capacity, in comparison to the Rzeppa joint that is described next. Also, the Weiss, like the Tracta joint, is not self-supporting, so that similar limitations apply as to the types of vehicle on which it may be used.

4.2.4 Rzeppa joint

The operating principle of this type of ball constant-velocity joint was devised in 1926 by the Czechoslovakian inventor A H Rzeppa (pronounced 'Sheppa'), a one-time consultant to an American firm of gear-grinding machine manufacturers. With further development the Rzeppa concept has probably become the most widely-used design of constant-velocity joint in motor vehicle drive lines. (*Figure 4.7*).

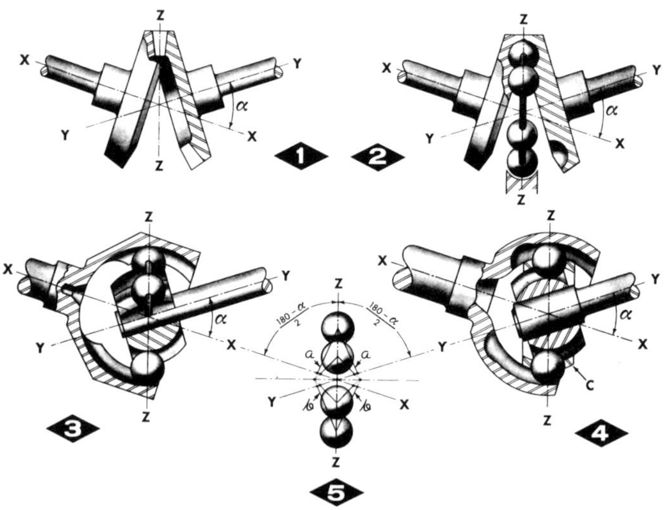

1 *Constant velocity bevel gear* (fixed angle)
2 *Constant velocity gear* (fixed angle). Driving balls replace gear teeth and mesh with pockets in gears
3 *Constant velocity joint* (variable angle). Transverse grooves in driving/driven members replace pockets
4 *Rzeppa CONstant VELocity joint.* Driving balls are engaged in transverse grooves and maintained in an angle bisecting plane Z-Z by ball cage C
5 *Constant velocity principle.* Driving balls move in a plane bisecting the angle included between shaft axes Z-Z. Every ball in every position maintains equal distance from both axes X and Y

Figure 4.7 Constant-velocity principle of the Rzeppa-type joint (*Con-Vel Dana*)

In basic construction it comprises a ball-and-socket joint with both parts grooved to embrace six rolling steel balls, which are held in a common plane by a controlling ball cage. The latter assembly thus constitutes the intermediate member of the joint and must be steered so that the six balls always lie in a plane which bisects the angle between the shafts, this being an essential condition for a constant-velocity drive. It is the manner of steering the ball-cage that is one of two features which distinguish the early from the later designs of Rzeppa joint. As originally designed, the ball-cage was steered by the toggling action of a 'pilot pin', the ball ends of which registered in the ends of the driving and driven shafts with endwise movements of the pin being allowed only in the latter.

In later designs, the ball cage was 'track' steered by the ingenious expedient of offsetting the geometric centres of the ball-and-socket grooves by equal amounts from the true centre of the joint. Whenever there is articulation of the driving and driven shafts, the deliberate mismatch in curvatures of the ball-and-socket grooves creates a wedging effect on the caged balls, until they retreat into a plane which again bisects the angle between the shafts. The same effect has been produced in a more recent development of the Rzeppa joint by offsetting the geometric centres of the cage inner and outer surfaces, so that the balls are 'cage' rather than 'track' steered. This in turn allows the use of straight instead of curved ball grooves in the socket housing, so that a certain amount of plunging as well as complete universal action of the drive shaft can be accommodated within the joint itself, thereby eliminating any need for a splined coupling. Track- and cage-steered Rzeppa type joints are therefore sometimes also referred to as 'fixed' and 'plunging' joints respectively (*Figure 4.8*).

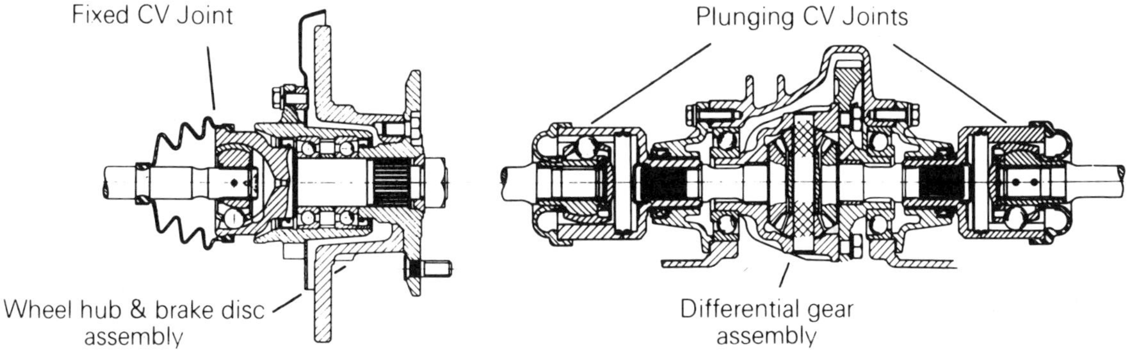

Figure 4.8 Typical application of Rzeppa-type 'fixed' and 'plunging' joints (*GKN Hardy Spicer*)

The second feature which distinguishes the early from the later designs of Rzeppa joint is related to the cross-sectional shape of the ball grooves. Originally these were of purely circular form, so that torque was transmitted with the balls loaded in shear. This could lead to severe edge loading and wearing of the grooves, which then resulted in noisy operation of the joint. To overcome this problem Hardy-Spicer in this country have long since modified the form of the grooves to an elliptical arc cross-section, so that the balls not only assume an angular contact away from the edges of the grooves, but also become advantageously loaded in compression rather than shear. The Hardy-Spicer version of Rzeppa joint incorporating this feature and also a track-steered ball-cage were introduced some 20 years ago on the Austin-Morris Mini cars. In modern form, the Rzeppa type of joint offers the several advantages of being structurally self-supporting and compact in relation to its torque capacity, and having the ability to operate quietly at high joint angles up to about 45°.

4.2.5 Lubrication and sealing of constant-velocity joints

The double-Hooke's, Weiss and Rzeppa types of constant-velocity joint are generally grease-lubricated for life and sealed by a synthetic rubber sealing boot when used on the drive shafts for independently-sprung wheels. This form of lubrication and sealing is not required when Tracta, Weiss and Rzeppa joints are installed in live axles with steerable wheels, because part-spherical enclosures are provided at the steering pivots to seal them and they can receive their lubrication from the same source as that used for the final-drive gears.

4.3 RUBBER-ELEMENT UNIVERSAL JOINTS

Since the early history of the motor car automobile engineers have always been attracted to the idea of a universal joint that does not require lubrication or any other attention during the useful life of the vehicle. In practical terms, this led to the development of various forms of universal joint in which the driving and driven shafts were provided with spider-arm connections to a flexible disc of either metal or leather. It was the British engineer E J Hardy who, in 1911, introduced the first really successful flexible disc universal joint. The disc was built up from many layers of rubberised fabric material, the weave of successive layers being staggered to provide a uniform strength of disc. Later developments in this field became concentrated on using rubber as the flexible medium in the form of either bushings or as a complete ring.

Before describing the various modern types of rubber element universal joint, it may be helpful to list the looked-for advantages in using such joints, as follows:

(1) Eliminate lubrication and wearing problems in the propeller-shaft assembly.
(2) Reduce transmisison of vibration and noise.
(3) Cushion shock loads on the transmission system.

On the debit side, rubber element universal joints generally do not provide a constant-velocity drive when operated at an angle.

4.3.1 Layrub joint

This represents a very well-known version of the ‘rubber-bushed’ type of universal joint, which was once widely used on the propeller shafts of commercial vehicles and, rather interestingly, on electrically-driven trolleybuses to cushion their suddenly-applied and high starting torque.

It had no ‘working parts’ in the accepted sense of the term, because all relative movement in the joint was accommodated by means of rubber bushings housed between the complementary halves of a circular steel housing (*Figure 4.9(a)*). The rubber bushings were under compression in their housing and had specially-shaped cavities at their ends. Steel trunnion sleeves having spigoted shoulders were pressed into the rubber bushes. Each trunnion sleeve located in a registered portion of bolt hole in the two-armed spider flanges of the driving and driven shafts.

For the propeller-shaft speeds at which the Layrub universal joint was normally operated no centring device was required, because the effect of any imbalance of the propeller-shaft assembly would not be sufficient to cause whirling. Its angular capacity was generally limited to about 5° for continuous deflection. The axial flexibility of the rubber bushings was sufficient to make redundant the need for a sliding-spline joint in the propeller-shaft assembly.

4.3.2 Rotoflex joint

For this version of ‘rubber-ring’ or ‘doughnut’ type of universal

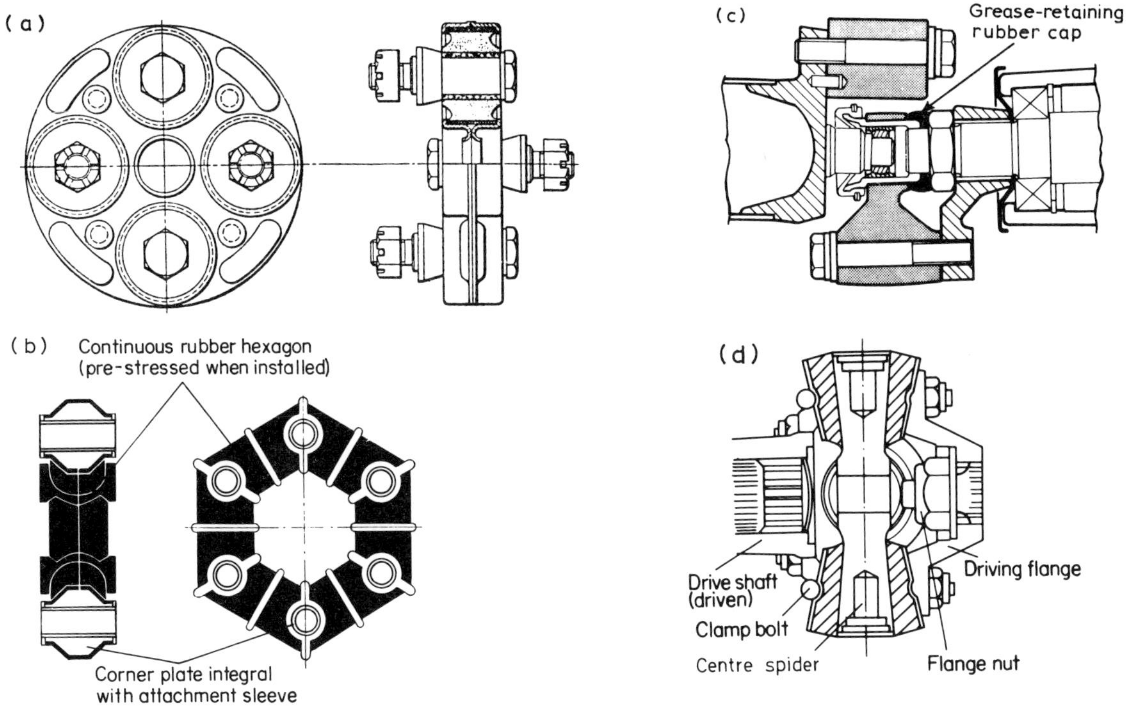

Figure 4.9 Types of rubber-element universal joint. (a) Layrub joint, (b) Rotoflex joint, (c) rubber-element joint with centring device (*Alfa Romeo*), (d) Moulton joint

joint, a regular polygon of rubber with metal bushes vulcanised at the corners is employed (*Figure 4.9(b)*). The driving and driven shafts are provided with three-armed spider flanges for bolt connections to the metal bushes. An important design feature of this type of universal joint is that the rubber element is supplied pre-compressed by an encircling steel band, which is then discarded after installation of the joint. Its purpose is to ensure maximum endurance of the rubber element in service, since the rubber remains in a compressed state during operation. For high-speed applications joints of this type require a mechanical centring device, the associated spherical seat and bush being pre-packed with grease and sealed with a protective cap (*Figure 4.9(c)*). The purpose of the mechanical centring device is, of course, to restrain any tendency towards whirling of the propeller-shaft assembly for which it is never possible to achieve perfect balance, as explained in § 4.8.2 of *Vehicle Technology 1*.

4.3.3 Moulton joint

This type of rubber-element joint (*Figure 4.9(d)*) is so-named because it was evolved and patented by Moulton Developments Limited in this country. It is basically of the Hooke's type but its four-armed steel spider carries bonded-on, part-conical, rubber bushes. These are embraced by split sleeves and retained against the saddle-shaped forked ends of the driving and driven shafts by U-bolts and nuts. The U-bolts are shouldered to ensure that the correct amount of pre-load is applied to the rubber bushes on assembly of the joint. In this context, it should be appreciated that the fatigue life of rubber is always improved if it is never allowed to relax completely when undergoing changes in loading—exactly the same principle being applied by the civil engineer to pre-stressed concrete!

4.4 POT TYPES OF UNIVERSAL AND SLIDING JOINT

These take two basic forms known as 'bi-pot' and 'tri-pot' joints (*Figure 4.10*), the latter being a modern development of the former, which in turn has its origins in the very early history of the motor car. Both forms of pot joint have the inherent attraction of being able to accommodate large endwise movements of the propeller or drive shaft while transmitting torque, thereby eliminating the need for a splined coupling. They are therefore said to possess a 'free-plunging capability'.

4.4.1 Bi-pot joint

This type of pot joint first appeared as long ago as 1894 on early steam cars and was revived later on the once well-known French De Dion cars. On these the rear wheels were carried by a 'dead' axle beam curving around behind the final drive, and were driven by two drive shafts each equipped with two universal joints. Since the arcuate movements of the short drive shafts mismatched those occurring at the axle-beam ends, each shaft together with its two joints had to be capable not only of universal, but also of telescopic action. This type of 'De Dion' drive is, in fact, still favoured on certain high-performance cars, because it eliminates the wheel-tramping tendencies of a live axle.

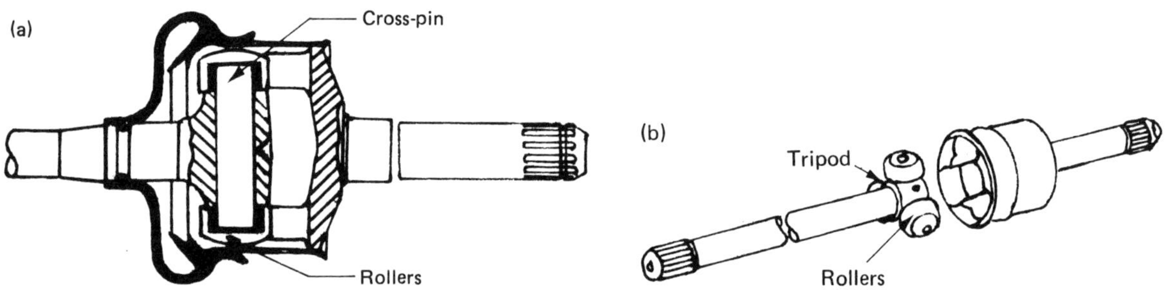

Figure 4.10 Comparison between bi-pot (a), and tri-pot (b), universal joints (*SAAB*)

Early constructions of bi-pot joint, usually referred to by motoring historians as 'De Dion' and 'sliding-block' joints, comprised simply a cross-pin at the end of the drive shaft, each end of the pin being free to pivot within its own square-shaped sliding block. These in turn contacted the flat faces of two slots in the joint housing. The sliding blocks were also provided with radiused ends to allow the shaft to rock in the plane of the cross-pin and therefore complete its universal action. Telescopic movement or 'plunge' within the joint was, of course, accomplished by the action of the blocks sliding endwise along the housing slots.

In modern versions of what is now classified as a bi-pot joint, the pivoting and sliding blocks are replaced by part-spherical rollers operating within partly cylindrical bores in the joint housing. The part-spherical rollers are themselves supported on the cross-pin by needle-roller bearings. Such joints (notably the 'Detroit' type) were once fairly widely used at each end of the normal propeller shaft on American cars and also on certain high-grade cars elsewhere, because they did of course eliminate the need for a sliding splined joint. It is now more usual to find the bi-pot joint at the inner ends of drive shafts for independently-sprung and driven rear or front wheels. This type of joint is not suitable for the outer ends of front-wheel drive shafts, because of its non-constant-velocity operation.

4.4.2 Tri-pot joint

As its designation would suggest, this type of pot joint features a tripod member comprising three equi-spaced radial pins and part-spherical rollers and may be regarded as a further development of the bi-pot joint. It was introduced in its present-day form by the French firm of Glaenzer-Spicer, who in the early nineteen-sixties called it the 'Tripode' universal joint. The tri-pot joint shares the same advantage as the bi-pot one in that it will accommodate endwise movement within the joint itself with a minimum of frictional resistance. Its principal advantage over the bi-pot joint is that it will transmit a very nearly constant-velocity drive, which perhaps calls for a little more explanation.

From the descriptions given in § 4.2 of the double-Hooke's, Tracta, Weiss and Rzeppa joints, it may be recalled that in all cases the intermediate member of the joint (such as the ball-cage in the Rzeppa joint) is constrained always to lie in a plane which bisects the angle between the driving and driven shafts. This plane is known in fact as the 'median' or 'homokinetic' plane and is an essential condition of operation to provide a constant-velocity drive. It is, therefore, somewhat difficult to visualise how the tri-pot joint without such an intermediate member can provide this, since the median plane always lies perpendicular to the axis of the drive shaft. In actual practice, however, the variations in angular velocity are largely compensated for by the tripod member weaving very slightly off its own centre three times per revolution, so that to all intents and purposes the tri-pot joint is a constant-velocity one. It has a good torque capacity for its diameter by virtue of the load being divided among three rolling elements. This feature, together with its simplicity of construction, has led to the wide usage of the tri-pot joint on the drive shafts of front-wheel-drive cars. The modern pot joint is sealed by a rubber bellows and requires no further lubrication after initial build.

4.5 FRONT-WHEEL DRIVE SHAFTS

Before commenting upon the construction of front-wheel drive shafts, it may be useful first to compare their duty with that of an equivalent single propeller shaft in the Hotchkiss-drive system, as described in § 4.1 of *Vehicle Technology 1.*

(1) The location of the drive shafts is such that they are necessarily shorter in length.
(2) Since the drive shafts operate directly between the final drive and the road wheels they transmit greater torque.
(3) From (2) it also follows that they rotate at a slower speed.
(4) When used in conjunction with inboard-mounted front brakes, the drive shafts must be able to withstand brake torque loading which can be applied more suddenly than that generated by the engine and transmission system.

Naturally the above considerations apply also to the drive shafts of independently-sprung and driven rear wheels.

4.5.1 Equal length front-wheel drive shafts

In all front-wheel-drive cars with fore-and-aft-mounted engines and transaxles, the drive shafts can obviously be made equal in length. For the modern small- to medium-sized FWD car they are produced from steel bar and may therefore be described as being of 'solid', as opposed to 'tubular', construction. Their diameter is typically in the region of 20 to 25 mm, and according to the type of universal joint used either one or both ends may be splined to provide an interference fitting for the inner member of the joint. A locking ring generally completes the assembly of shaft-end and joint. This is usually supplied as a replacement unit, because the relevant parts are individually matched and are therefore not interchangeable.

Tubular-construction drive shafts may be encountered where double-Hooke's joints are used, because this type of joint lends itself to easier attachment to a tubular shaft and may also require a sliding spline coupling. A shaft diameter of about 40 mm would be typical in such cases.

4.5.2 Unequal length front-wheel drive shafts

For front-wheel drive cars with transversely mounted engines and transmissions, the final-drive housing can be offset appreciably from the mid-track of the wheels, such that the difference in lengths between the long and short drive shafts can be typically in the ratio of 3:2. In practice this may cause a steering pull to one side on hard acceleration

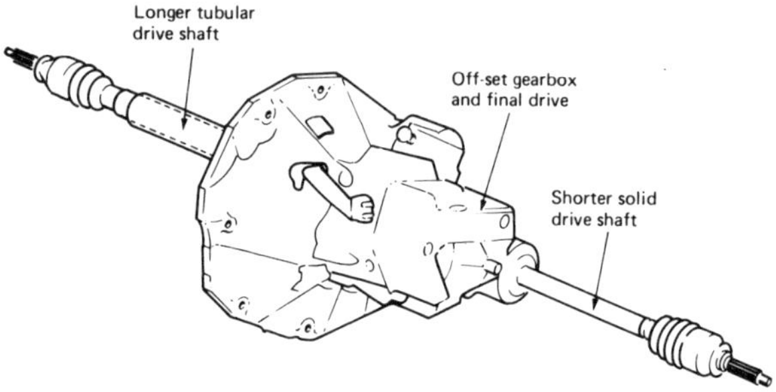

Figure 4.11 Tubular and solid FWD shafts (*Fiat*)

for the following reasons:

(1) If the long and short shafts are made the same diameter then the difference in their torsional rigidity results in the former taking longer to 'wind up' than the latter, so that power is supplied to one wheel momentarily before the other. To minimize this effect the longer shaft is stiffened by making it of larger diameter tubular construction, while the shorter shaft remains of smaller diameter solid construction (*Figure 4.11*).

(2) Any difference in the relative droop angles of the drive shafts as the front of the car rises on hard acceleration, implies that friction losses are increased in the univeral joints of the more steeply angled shorter shaft, so that slightly less power is supplied to that particular wheel. This can only be remedied in designs where a jackshaft and steady bearing are added to equalize the lengths of the drive shafts.

4.6 TORQUE-TUBE DRIVE

Although now rarely employed in its original form, an early rival to the Hotchkiss-drive system (and paradoxically later used by Hotchkiss themselves) was that known as the 'torque-tube' drive (*Figure 4.12*). In this system the revolving propeller shaft was totally enclosed within a rigid tube or 'torque tube', which at its front end usually pivoted about a hollow spherical bearing just behind the gearbox, and at its rear end was attached to the nose-piece of the axle. Adjustable diagonal radius rods were generally utilised to

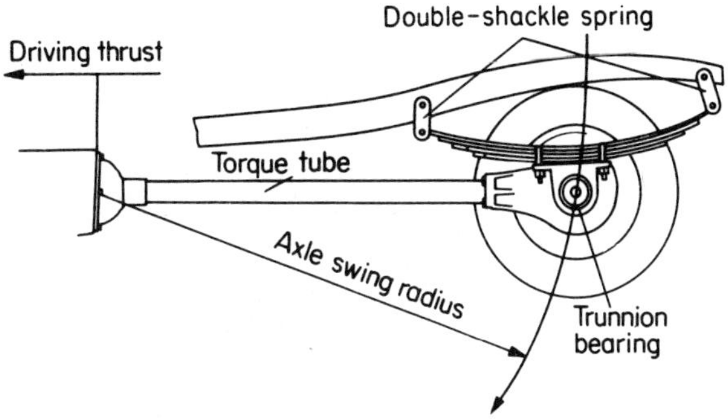

Figure 4.12 Layout of full torque-tube system

maintain the torque-tube in proper alignment with the rear axle. Since the relative movement between the rear axle and the gearbox was confined to a simple angular one, the inner propeller shaft required only a single universal joint at its front end.

Where semi-elliptic leaf springs were used for the rear suspension they had to be shackled at both ends, so as to accommodate the swinging motion of the torque tube and axle with rise and fall of the road wheels. Trunnion bearings were ideally, but not always in practice, provided for anchoring the axle to each spring. It should be appreciated that the fore-and-aft location of the axle was finally dependent upon the engine and gearbox mountings, since end-thrust loads due to wheel traction and braking were transmitted to them via the spherical bearing of the torque tube.

4.6.1 Semi-torque-tube drive

General Motors Opel in Germany introduced this compromise form of torque-tube drive in the early nineteen-sixties, since when there have been several other examples of its use.

It comprises an abbreviated-length torque tube, or 'pinion-housing extension', which is flange-mounted to the nose of the axle (*Figure 4.13*). The front end of the torque tube is pivoted from various forms of rubber bearing supported from the understructure of the car body. Within the front end of the torque tube is the equivalent of a propeller-shaft centre-bearing. This carries the input end of either a solid or tubular 'secondary drive shaft', the output end of which is connected to the final-drive bevel pinion shaft through a splined muff coupling. This comprises simply a cylindrical collar with internal splines that is pressed over the splined ends of the two co-

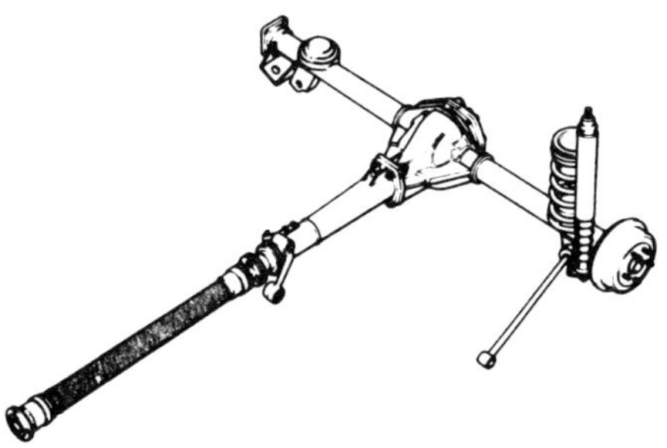

Figure 4.13 A modern semi-torque-tube system with plunger-type constant-velocity universal joints on primary propeller shaft (*GKN Hardy Spicer*)

axial shafts being connected, approximately one-half of the collar contacting the end of each shaft. In its various forms the muff coupling has its origins in very early engineering practice, since a common problem in mechanical design is that of rigidly connecting together two shafts.

A splined connection is also provided for a coupling flange at the front end of the secondary drive shaft, which in turn connects to the rear universal joint of an open primary propeller shaft. The latter is equipped with either two Hooke's universals and a sliding joint, or simply two constant-velocity joints of the plunging type, since a limited amount of fore-and-aft and vertical movements of the rubber-mounted torque tube and axle must be permitted relative to the power unit. To assist in maintaining alignment between the semi-torque tube and axle, rubber bushed longitudinal links pivot between the axle ends and brackets attached to the body structure. The radius of articulation of these links generally coincides with that of the axle and torque tube about its forward rubber bearing. To prevent sideways movement of the axle a separate means of control must be used, such as a Panhard rod described in § 6.17.

4.7 THE UTILISATION OF TORQUE-TUBE DRIVE

To appreciate why the 'full' torque-tube drive has long been obsolete and the 'semi' version continues to find limited acceptance, it is necessary to make a few general comparisons not only between these two forms of torque-tube drive, but also to relate them to the rival Hotchkiss-drive system as described in § 4.1 of *Vehicle Technology 1.*

4.7.1 Full torque-tube *v* Hotchkiss-drive

(1) The alternative full torque-tube drive immediately relieved the leaf springs of axle casing torque reaction, together with the thrust loads associated with traction and braking of the wheels.
(2) Since with torque-tube drive the springs are left to perform only their vehicle-supporting function they can be made more flexible to improve ride comfort.
(3) Similarly, the adoption of torque-tube drive allowed the early use of friction-free coil springs for non-independent rear suspension, again to improve ride.
(4) Full torque-tube drive required the use of only one universal joint in the drive-line, albeit at the time a non-constant-velocity one with its usual disadvantage.
(5) A major disadvantage of full torque-tube drive was that it increased the unsprung weight of the rear suspension system, which could adversely affect road-holding.
(6) The axle casing was subjected to greater bending loads in the horizontal plane, because of the traction and braking thrust loads being transmitted to the vehicle mainly by the central torque tube.
(7) Undesirable transmission noise could be telegraphed to the car interior by the hollow torque tube, which thus acted as a sound-amplifier.
(8) The full torque-tube drive generally demanded a more elaborate overhaul procedure with consequent higher repair costs.

4.7.2 Semi- *v* full torque-tube drive

(1) In comprising little more than a pinion-housing extension, the semi torque-tube drive is at a much less disadvantage in terms of rear-suspension unsprung weight.
(2) By virtue of its rubber-mounting system the semi-torque tube with separate locating links provides better isolation from road noise and transmission vibrations.

(3) The semi-torque tube permits a reduced height of transmission tunnel in the body floor structure, thereby improving rear-seat passenger accommodation.
(4) For additional refinement the short primary propeller shaft can be of one-piece construction and embody modern constant-velocity joints of the plunging type.
(5) In reacting crown-wheel torque (not 'pinion torque' as is sometimes inferred) the semi-torque tube is of a more convenient length to act as a short lever arm and reduce rear-end squat of the car during acceleration. Apart from this 'anti-squat' effect it likewise confers an 'anti-dive' effect when the axle casing is resisting brake torque reaction. Pitching of the car during these manoeuvres was illustrated in *Figure 1.11* of *Vehicle Technology 1.*
(6) The semi-torque tube still remains more complicated than a Hotchkiss-drive, but is less complicated than the driving arrangements for independently-sprung rear wheels with which system of suspension it is has often been favourably compared.

4.8 ARRANGEMENT OF PROPELLER SHAFTS WITH FOUR-WHEEL DRIVE

The purpose, construction and operation of the transfer gearbox that is either attached directly, or coupled separately, to the main gearbox has already been explained in §§ 3.9 to 3.11, so it now remains to complete the picture by describing the propeller-shaft arrangements (*Figure 4.14*).

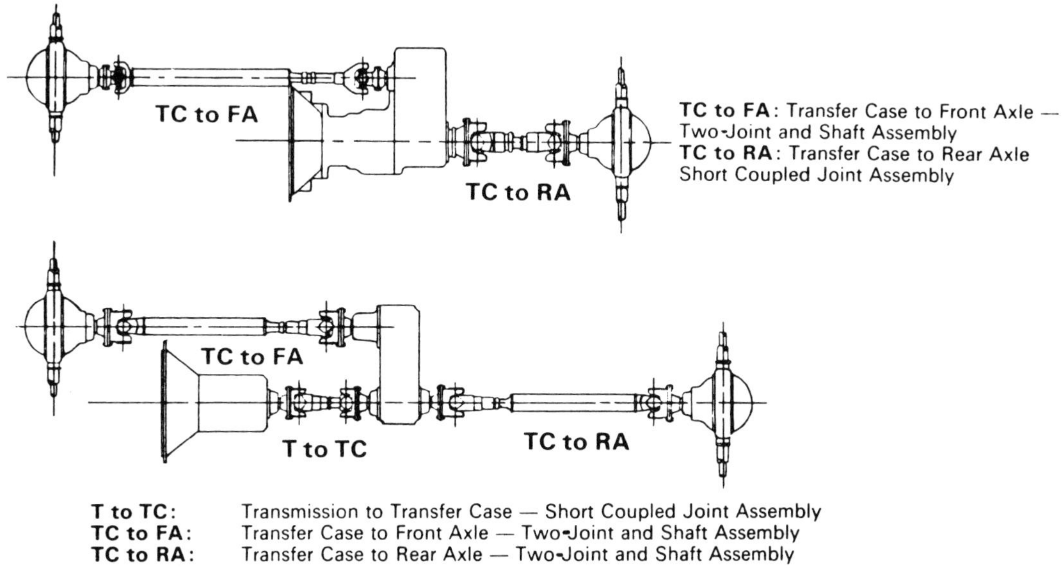

Figure 4.14 Conventional arrangements of propeller shafts used with 4WD (*GKN Hardy Spicer*)

In the conventional type of cross-country vehicle, the drive from the transfer box is transmitted to the front and rear final-drive assemblies through two propeller shafts, each with Hooke's-type universal joints at each end and incorporating a sliding coupling. The rear propeller shaft is typically coupled to a live rear axle in the conventional manner. In the case of the front propeller shaft, which runs alongside the power unit, this again may be coupled to a live

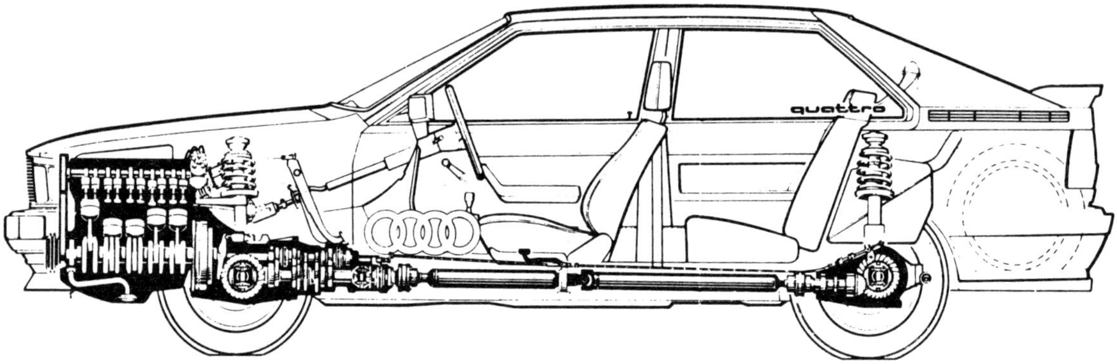

Figure 4.15 A modern 4WD arrangement of front transaxle and rear propeller shafts (*Audi*)

axle or, less commonly, to a final-drive unit mounted on the vehicle structure and driving independently-sprung front wheels via the usual universally-jointed drive shafts.

A more recent innovation has been to modify the transaxle of a basically front-wheel-drive car with the object of providing an offtake for additional rear-wheel drive (*Figure 4.15*). When this is done there is, of course, no need for a separate front propeller shaft and the rear one can be of either single or divided construction for coupling to a structure-mounted final drive. The latter is required for the independently-sprung rear wheels of cars with this very modern application of four-wheel drive.

5 Final drive

5.1 FINAL-DRIVE GEARS AND BEARINGS

It may be recalled that the various types of bevel final-drive gears, together with their advantages and disadvantages, were examined in § 5.2 of *Vehicle Technology 1*. However, it may be useful to summarise the basic requirements for the durable and quiet operation of these highly-stressed gears, as follows:

(1) Accuracy and consistency in the manufacture of the gears themselves.
(2) Rigidity in mounting of the gears as provided by their housing and bearings.
(3) Correct choice and application of axle lubricant.

Although we are not concerned here with the specialised design and manufacturing processes associated with the final-drive gears themselves, it is necessary for the service engineer to gain an understanding of their bearing mountings, adjustment and lubrication requirements.

5.1.1 Crown-wheel-and-pinion bearing loads

The loads acting on the crown-wheel-and-pinion bearings are primarily those which result from the meshing action of the gear teeth and are most conveniently presented by illustration, as in *Figure 5.1*. Here, in the case of spiral bevel gears, it will be seen that several

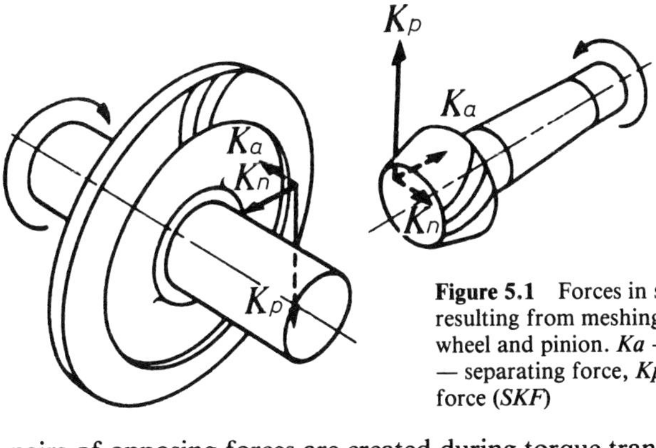

Figure 5.1 Forces in spiral bevel gears resulting from meshing action of crown wheel and pinion. *Ka* — thrust force, *Kn* — separating force, *Kp* — tangential force (*SKF*)

pairs of opposing forces are created during torque transmission, as below:

(1) An end-thrust force on the pinion with a corresponding radial separating force on the crown wheel.
(2) An end-thrust force on the crown wheel with a corresponding radial separating force on the pinion.
(3) The corresponding tangential forces on the crown wheel and pinion.

Whereas each pair of opposing forces is equal in magnitude with spiral-bevel gears, this is not so in the case of the relatively more complicated meshing action of hypoid gear teeth where the tangential forces responsible for torque transmission are neither opposite in direction nor equal in magnitude. It should be noted that when the engine is being used as an aid to braking on vehicle overrun, the loads acting on the gears are reversed in direction and similarly, of course, when reverse gear is engaged.

A point of interest in connection with spiral-bevel and hypoid gears is that, unlike straight-toothed bevel gears, the curvature of their teeth produces a 'winding-in' or 'winding-out' action. The resulting end-thrust on the pinion must therefore be taken into account, along with that arising purely from its conical form. In fact, the high loading imposed on the pinion bearings makes their application the most critical in the axle or its equivalent.

The additional loads that may be imposed on the crown-wheel-and-pinion bearings, at least so far as a live axle is concerned, are respectively radial ones from the inner ends of semi- and to a lesser extent three quarter-floating axle shafts, and axial ones from any sliding-joint friction in the propeller shaft.

5.1.2 Crown-wheel-and-pinion bearings

It is generally regarded as being less difficult to maintain spiral-bevel and hypoid gears in satisfactory alignment with rolling-element bearings, than it would be with plain bearings. The use of rolling-element bearings, especially those of the tapered-roller type, has therefore long been established practice for supporting the final-drive gears of motor vehicles. Before describing the various arrangements of crown-wheel-and-pinion bearings, it may be of interest to consider the principal advantages of using tapered-roller bearings, as follows:

(1) They possess true rolling characteristics because the projected conical surfaces of the rollers and their raceways all meet on the axis of the bearing. This is in contrast to the combined rolling and sliding that occurs between ball bearings and their raceway surfaces owing to the variations in circumferential speed of a ball rolling freely in a conforming groove. (The sliding effect is sometimes referred to as 'Heathcote slip', so-named after H L Heathcote who identified it in the early nineteen-twenties.)

(2) By virtue of their unique geometry, the alignment of the rollers is not wholly dependent upon either the thrust rib of the cone (the inner race of a tapered-roller bearing is termed

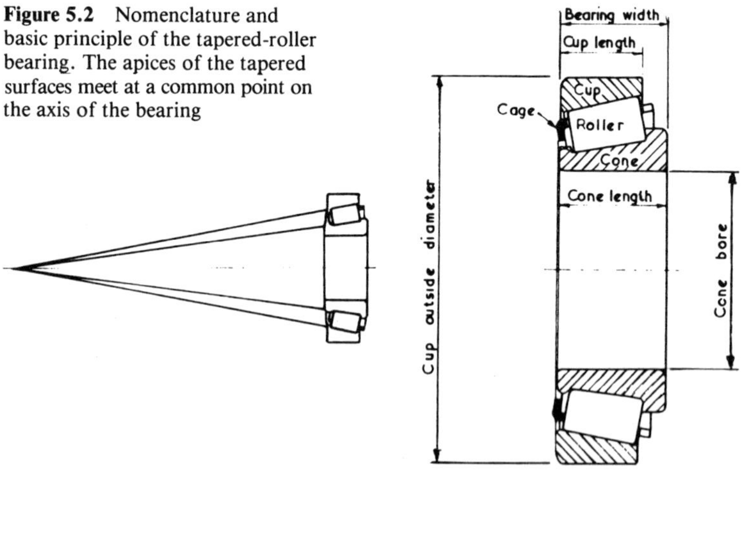

Figure 5.2 Nomenclature and basic principle of the tapered-roller bearing. The apices of the tapered surfaces meet at a common point on the axis of the bearing

the 'cone' and the outer race the 'cup'), or the bearing cage.

(3) The 'line contact' of the tapered-rolling elements, as opposed to the (theoretical) 'point contact' in ball bearings, provides a greater load-carrying capacity for a given size of bearing. This is of particular advantage in the case of the heavily-loaded pinion bearings, where there are space limitations on their size.

(4) Tapered-roller bearings are inherently rigid and have the ability to withstand any combination of radial, thrust and tilting loads within the designed capacity of the bearing. Shallow-angle bearings are specified for carrying heavy radial loads and steeper-angle ones for carrying heavier thrust loads.

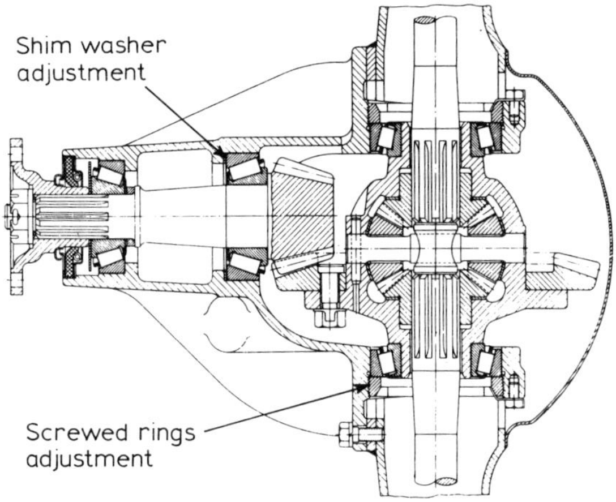

Figure 5.3 Tapered-roller bearing mounting for final-drive gears in banjo-type axle (*SKF*)

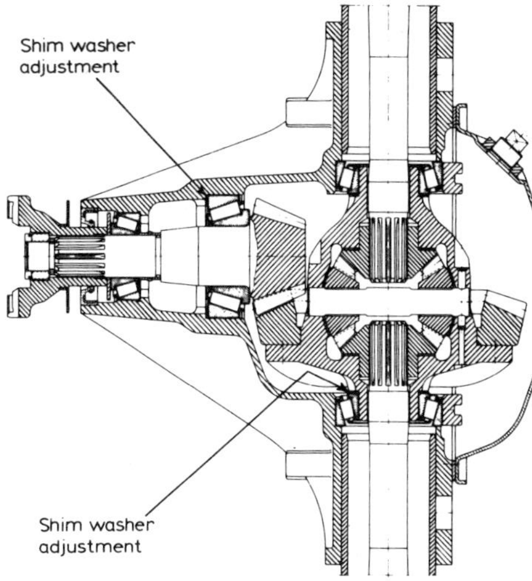

Figure 5.4 Tapered-roller bearing for final-drive gears in carrier-type axle (*SKF*)

(5) Due to their tapered construction, the operating geometry remains undisturbed when bearings of this type are adjusted to remove initial play and also preloaded as described at a later stage.

(6) Although the ultimate fate of all types of rolling-element bearing is failure by fatigue, in practice the useful life of the final-drive bearings is usually determined by wearing of the pinion bearings causing noisy operation of the gears. In this context, the rate of wear for tapered-roller bearings is generally lower than that for ball bearings.

(7) Tapered-roller bearings possess the useful property of acting as their own oil pump, since by virtue of their form oil flows through the bearing from the small end to the large end of the rollers.

5.1.3 Crown-wheel-and-pinion bearing arrangements

In modern automotive practice, the bearings for the crown wheel and pinion may be mounted in either a live rear axle, a separate final-drive housing attached to the vehicle structure, or a combined gearbox and final-drive assembly known as a 'transaxle'.

For live rear axles and separately-mounted final drive (*Figures 5.3, 5.4, 5.10*), the arrangement of the crown wheel and pinion bearings is typically as follows:

The pinion: This is made integral with its shaft and is overhung from two tapered-roller bearings mounted 'back-to-back' (small ends of rollers pointing inwards). These are spaced apart by their outer races or cups, which locate against shoulders in the pinion housing. Endwise retention of the pinion shaft is effected through the inner races or cones of the bearings, since these are rigidly held between the pinion and the boss of the companion flange for the rear univeral joint of the propeller shaft by a clamping nut on the front end of the shaft.

The bearing cups are pressed into the pinion housing, similarly the cone of the bearing adjacent to the pinion is made an interference fit on the shaft. A lighter fit suffices for the cone of the other bearing at the less heavily-loaded end of the pinion shaft, since this facilitates the build and adjustment of the pinion and bearings. The spacing of the pinion bearings is generally 2½ times the overhang distance of the pinion.

The crown wheel: This is bolted behind its mesh to the flanged differential cage, which in turn is carried by two tapered-roller bearings mounted 'face-to-face' (small ends of the rollers pointing outwards) in the final-drive housing. On each side of the differential cage bosses are machined to locate the bearing cones, these being made an interference fit on their seatings. For endwise retention of the crown wheel, the cups of the bearings locate against shouldered bores in the final-drive housing, in which they are made a lighter fit, again to facilitate build and adjustment. Detachable bearing caps are generally required to permit assembly of the crown wheel and bearings into the final-drive housing. The spacing of the crown wheel bearings usually corresponds to about three-quarters of the mean pitch diameter of the crown wheel.

In heavy-duty rear axles, the pinion may be 'straddle'-mounted by having a pair of back-to-back tapered-roller bearings in front of it and a single parallel-roller bearing immediately behind it (*Figure 5.5*). The purpose of this arrangement is, of course, to increase the rigidity of mounting for the pinion. Since the crown wheel must likewise be immune from deflection under heavy side loading, it is often provided with a 'steady pad' that is mounted from the final-drive housing. This steady pad is either screw- or shim-adjusted to be just

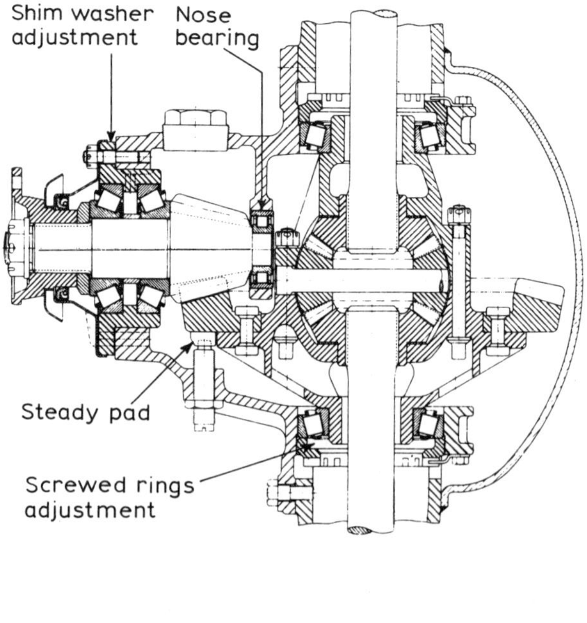

Figure 5.5 Straddle-mounted pinion in heavy-duty axle (*SKF*)

clear of the non-meshing face of the crown wheel under light-load running conditions.

Where the final drive is combined with the gearbox to provide a transaxle (*Figure 5.6*), the bearing arrangements for the pinion may differ slightly from those found in the live axle, because the gearbox layshaft also serves as the bevel pinion shaft as explained in § 3.2.7 of *Vehicle Technology 1.* The pinion-cum-layshaft is therefore supported from the gearbox part of the transaxle housing usually by either two very widely spaced tapered-roller bearings mounted back-to-back, or a single parallel-roller bearing at the pinion end of the shaft and a pair of back-to-back tapered-roller bearings at the other end.

5.2 METHODS OF ADJUSTING FINAL-DRIVE GEARS

An important consideration in the mounting of final-drive spiral-bevel and hypoid gears is that convenient axial adjustment of position is necessary for both the pinion and the crown wheel. At this stage a knowledge of the adjustment sequence, desirable tooth-contact patterns, and correct backlash between the gear teeth is not required, but rather the mechanical means by which these adjustments can be made.

5.2.1 Pinion depth adjustment

Axial positioning of the pinion may be obtained by either of the following methods:

(1) The insertion of an adjusting shim washer of appropriate thickness between the inner bearing cup and its locating shoulder in the housing (*Figures 5.3, 5.4*).
(2) The insertion of an adjusting shim washer of appropriate thickness between the inner bearing cone and the adjacent shoulder of the bevel pinion.

5.2.2 Crown wheel lateral adjustment

Axial positioning of the crown wheel may be obtained by either of the following methods:

(1) The transference of an adjusting shim washer of appropriate thickness from behind the bearing cone on one side of the differential cage to behind the bearing cone on the other side (*Figure 5.4*).
(2) The equal screwing and unscrewing of threaded sleeves that bear against the outer ends of the bearing cups (*Figure 5.3*).

5.3 BEVEL-GEAR DIFFERENTIAL

The need for a differential mechanism in the final drive of a motor-vehicle and the manner in which it behaves was explained in § 5.3 of *Vehicle Technology 1.*

Historically, the invention of the differential is usually attributed to the Frenchman Onésiphore Pecqueur, who in 1828 patented a steam vehicle that included a differential gear on the driving axle. However, a student of mechanics may well argue that the motor-vehicle 'differential gear' is technically not a differential mechanism at all, but a 'cyclic equalising gear'. This is because its basic function is that of dividing the crown-wheel torque equally between the two driven wheels, irrespective of their relative speeds of rotation. A true differential mechanism, on the other hand, is characterised by similar parts moving simultaneously at different rates. An example of this is the familiar differential chain hoist in which the endless chain operates over a compound sprocket-wheel having two different radii.

Be that as it may, perhaps we should stick to convention and continue to call it a differential gear (or even 'diff-unit' as it is universally called in motor-vehicle repair work).

5.3.1 Construction of the differential gear

We now have to consider the mechanical construction of the most usual form of differential used in motor vehicles, which is known as the 'straight-tooth-bevel' type and comprises the following parts:

Differential cage: From § 5.1.3 describing crown-wheel bearing arrangements, it may be recalled that the crown wheel is bolted to the flanged differential cage, which in turn is carried by the crown-wheel bearings. To gain maximum rigidity the differential cage is provided

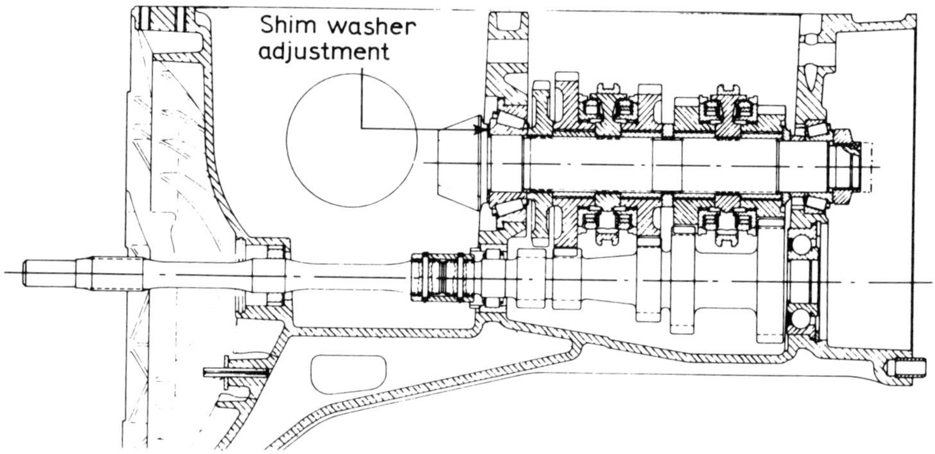

Figure 5.6 Typical mounting of bevel pinion in a transaxle (*Alfa Romeo*)

with substantial stiffening ribs and is produced from a malleable iron casting. It may be of either one- or two-piece construction, according to whether it contains two or four 'planet' bevel pinions. The latter number is used in heavy-duty axles so that contact pressure between the differential gear teeth is reduced by distribution (*Figure 5.7*).

Planet pinions: These bevel pinions are produced from steel and provided with coarse-pitched teeth of stub form. They are free to rotate on either a shaft located and retained by a tapered pin in the cross-bore of a one-piece cage or, in the case of a four-planet pinion design, on a spider-shaped member that is embraced by a two-piece

1 Crown wheel
2 Flanged differential case
3 Bolt
4 Nut
5 Flat thrust washer
6 L/h sun pinion
7 Planet pinion
8 Part-spherical thrust washer
9 Spider
10 R/h sun pinion
11 Flat thrust washer
12 Plain differential
13 Capscrew
14 Differential bearing race
15 Differential bearing track

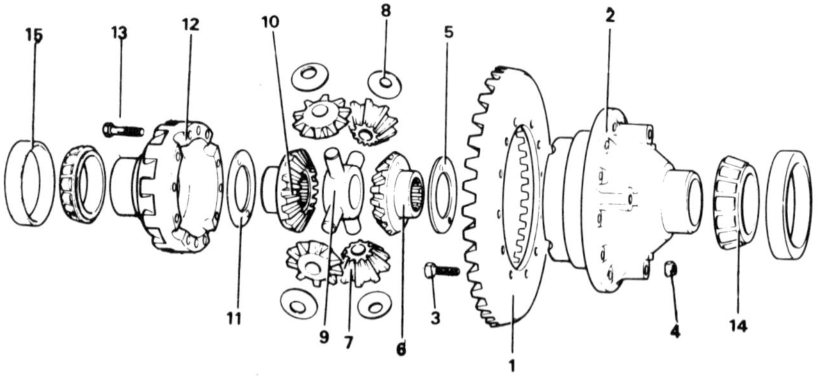

Figure 5.7 Exploded view of a heavy-duty four-planet pinion differential (*Seddon Atkinson*)

cage. The thrust faces of the planet pinions are of part-spherical form, as are their thrust washers and seatings in the differential cage. *Sun pinions:* Also known as 'side pinions', these are similarly bevel-toothed and have internally-splined hubs registering in recesses in side bores of the differential cage. Their thrust faces and those of the thrust washers and cage recesses are flat, since they can be made larger in diameter than the planet pinions and hence have sufficient bearing area to accept end-thrust loading. The inner splined ends of the axle half-shafts engage with the hubs of the sun pinions.

5.3.2 Operation of the bevel-gear differential

This is perhaps best explained by reference to the 'balance beam and slotted discs' analogy, in which the former corresponds to one of the planet pinions (input from the crown wheel) and the latter to the sun pinions (output to the axle shafts).

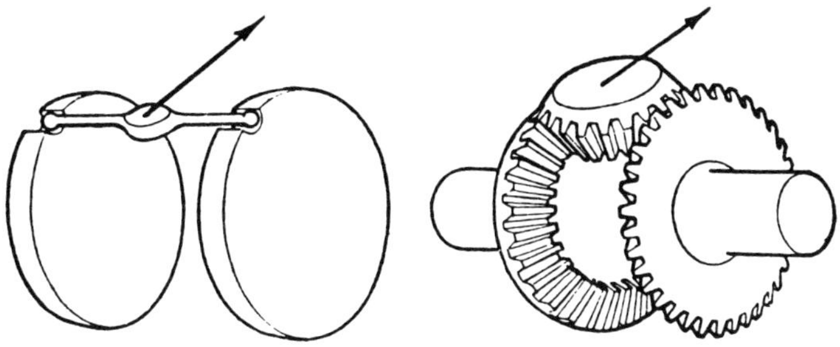

Figure 5.8 Operating principle of the bevel-gear differential

From this it should be evident that with the balance beam centrally pivoted, the loads transmitted by its ends to the rotatable slotted discs will be equal within the available swing of the beam. Such a differential would, of course, permit only a limited relative movement of the two discs, but if we could arrange for a whole series of balance beams extending to a full circle and then provide the discs with a complete set of slots, a continuous relative movement of the two discs can be accommodated. Here we have the essential ingredients of the real-life bevel-gear differential, which thus divides equally the turning force imparted by the crown wheel to the axle half-shafts, even though circumstances may dictate that one must rotate faster than the crown wheel and the other an equal amount slower.

It follows, of course, that if both half-shafts are rotating at the same speed as the crown wheel, there will be no relative movement between the planet and the sun pinions and the differential gearing and cage will simply act as a solid piece of metal.

5.4 LUBRICATION AND SEALING OF THE FINAL-DRIVE UNIT

It is perhaps a tribute to the well-planned lubrication and sealing of the modern rear axle that ultimately it wears generally rather than locally. In § 5.2.2 of *Vehicle Technology 1*, the special lubrication requirements of hypoid gears was mentioned, and it is now necessary to consider first the mode of application and then the sealing arrangements.

'Oil-bath' lubrication is employed for motor-vehicle final drives, the axle housing forming an oil-tight reservoir for the lubricating oil. A drain plug is usually provided on the underside of the housing and a combined filler and level plug on the rear cover. The oil level is generally made just high enough to cover the full-face width of the crown-wheel teeth at bottom dead centre.

5.4.1 Lubrication of the gears

At comparatively low road speeds this relies on the transport of oil by the crown-wheel gear, as teeth that have dipped into the oil travel round to the point of mesh with the bevel pinion. With increasing road speed oil splash or spray begins to be thrown up, which also serves to lubricate the differential gear assembly.

5.4.2 Lubrication of the bearings

Whereas the crown-wheel bearings may be splash-lubricated, the chief concern is to ensure that the heavily-loaded pinion bearings receive adequate lubrication under all running conditions. This is because under low-speed operation there is a tendency for the oil to cling to, and build-up on, the rim of the crown wheel so that oil in the reservoir can be drawn away from the rear of the housing to the detriment of pinion-bearing lubrication at the front end.

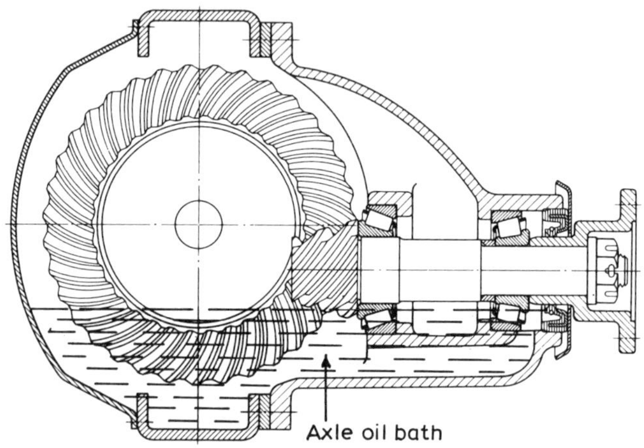

Figure 5.9 Lubrication of final-drive gears and pinion bearings (*SKF*)

As the road speed rises, however, there is an increasing tendency for the oil to be centrifuged from the rim of the crown wheel and thrown against the housing wall. This oil can then be collected to lubricate the pinion bearings by directing it between them via a suitably-positioned passage. Once it arrives between the bearings their tapered form enables them to act as their own oil pump to circulate the oil over the rolling elements. Research work carried out by the SKF Laboratory at Luton has shown that for the most effective lubrication of the pinion bearings their oil supply passage should be aligned with the rim, rather than the teeth, of the crown wheel.

5.4.3 Final-drive oil-sealing arrangements

The purpose of these is not only to seal the lubricant in, but also to prevent any foreign matter from entering. In modern practice, the sealing arrangements must take into account final drives incorporated in live rear axles, those separately-mounted on the car structure and others that form part of transaxles.

Live rear axles: Here we may confine our attention to the sealing of the bevel-pinion shaft, since the hub sealing at the axle ends was dealt with in § 5.4 of *Vehicle Technology 1.* An oil-sealing arrangement for the bevel-pinion shaft typically comprises a combination of the following:

(i) *A lipped metal dust shield* pressed against a shoulder on the companion flange and working in conjunction with the lipped casing of the oil seal.

(ii) *A radial-lip oil seal* pressed in a recess in the pinion housing and acting against a sealing surface machined on the boss of the companion flange. In this context, it should be appreciated that for proper functioning and satisfactory life of the seal, a very thin film of oil must always be maintained between its lip and the boss, the film thickness being controlled by the force exerted by the lip on the sealing surface. A certain tolerable leakage of oil therefore occurs, even though it should be hardly measurable in service.

(iii) *An oil slinger washer* sandwiched between the rear end of the companion flange boss and the pinion front bearing cone. Its purpose is to prevent flooding of the radial lip oil seal at high rates of oil circulation to the pinion bearings. However, an oil-slinger washer is not always included.

Separately-mounted final drives: The sealing arrangements for these are twofold, as follows:

(i) *Bevel-pinion shaft:* The arrangements for sealing this are identical to those previously described for the live axle.
(ii) *Stub drive shafts:* These connect with the differential sun pinions and are usually carried in rolling bearings mounted independently of those supporting the crown wheel and differential unit. A need is therefore created to seal the stub shafts where they emerge from the sides of the final drive housing. For this purpose radial lip seals are

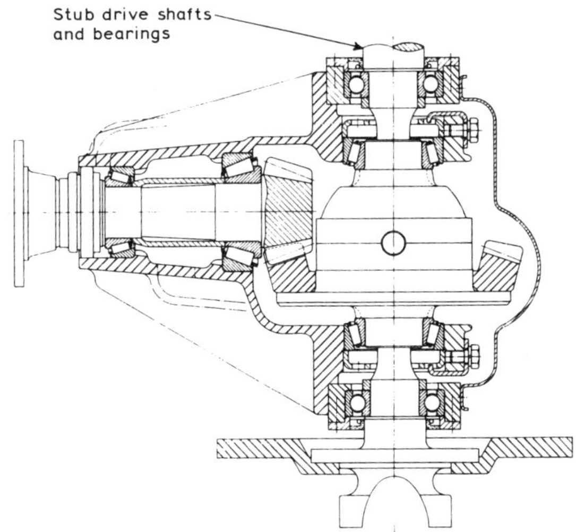

Figure 5.10 Arrangement of separately-mounted final-drive gears (*SKF*)

pressed into recesses machined in the separate bearing housings of the stub shafts and act against sealing surfaces provided near the flanged ends of the shafts. An additional sealing consideration arises where inboard brakes are located on each side of a separately-mounted final drive, since the stub shaft seals then have to be capable of withstanding limited periods of very high temperature operation.

Transaxles: The sealing arrangements for the stub drive shafts are identical to those previously described for the separately-mounted final drive. There is no sealing requirement for the bevel pinion-cum-layshaft, but as mentioned in § 3.8.2 of *Vehicle Technology 1,* a radial lip seal is required where the primary shaft enters the final drive portion of the transaxle casing. Its purpose, of course, is to prevent oil escaping from the transaxle into the clutch housing.

5.5 AXLE-CASING CONSTRUCTION

Before examining the different types of casing construction for live axles, it is necessary to identify the two main requirements that an axle casing must satisfy as a load-bearing structure, these being as follows:

(1) It must be sufficiently rigid to avoid any distortion of the final-drive housing under the most arduous operating conditions and thereby ensure quiet running of the gears.
(2) Its rigidity must also be such as to maintain the half-shafts in true alignment with the final-drive unit and thus reduce any possibility of their fatigue failure due to bending.

There are three basic types of live axle construction, two of which by virtue of their form (and not because of any lack in quietness of running!) have rather musical designations, as below:

(1) Trumpet
(2) Banjo
(3) Carrier

5.5.1 Trumpet axle

This is constructed from two trumpet-shaped axle halves, which are spigoted and bolted either directly together across their vertical joint faces, or indirectly together with an interposed aluminium alloy centre-section. In the former case one of the axle halves is extended beyond the true central parting line of the axle in order to accommodate the bevel pinion and its bearing mountings, while in the latter arrangement both the crown wheel and pinion may be supported independently of the axle halves by the centre-section housing.

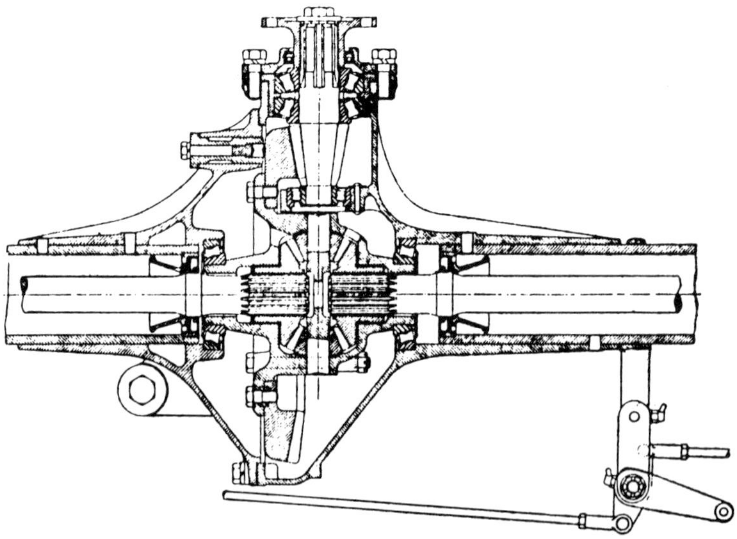

Figure 5.11 An interesting early post-war trumpet-type axle fitted to a commercial vehicle

The trumpet axle can enable the final-drive gears to be very rigidly supported and generally provides an unusually robust form of construction. It has, however, long been regarded as being somewhat difficult to manufacture and service, so that in recent years its application has been limited to a very few high-grade cars.

5.5.2 Banjo axle

For normal duty axles this type of casing is of one-piece construction. It is formed from a single tube and is provided with an enlarged open centre-section, which accommodates the flange-mounted final drive housing. The rear part of the centre-section is closed by a domed rear cover that need not be made detachable and can thus be welded onto the casing in the interests of both rigidity and oil tightness.

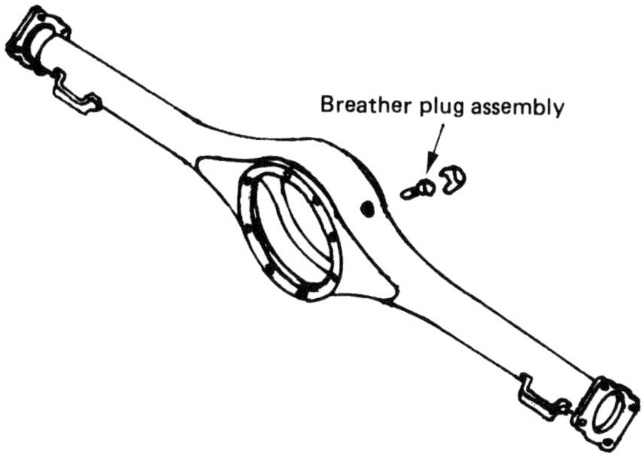

Figure 5.12 Banjo-type axle casing

Although this type of axle construction is, by virtue of its form, well-adapted to resist vertical bending loads, it is less able to resist tractive and braking forces that act in the horizontal plane. Also, the crown wheel tends to be less well supported, because its bearings are cantilever-mounted (overhung) from the final-drive housing (*Figure 5.3*). However, this type of axle construction does allow the final-drive assembly to become a self-contained unit, which was an important servicing consideration when axles were not so reliable as they are nowadays.

5.5.3 Carrier axle

The essential feature of this type of axle casing (*Figure 5.13*) is that the crown wheel and pinion are mounted in a rigid central housing or 'carrier' of malleable cast iron, into which the steel axle tubes are pressed and welded in position. This form of construction necessarily requires the crown wheel and differential unit to be installed through a large aperture in the rear of the carrier. Semi-circular shouldered recesses are machined in the carrier to locate the bearing cups of the crown wheel and differential unit, which are retained by bolted-on bearing caps. A detachable cover is, of course, provided to close the rear aperture of the carrier.

Where a live rear axle is still retained in the modern car, the carrier-type axle may be regarded as established practice. In this country it began to find increasing acceptance in the nineteen-fifties, but its American origins date back to the early nineteen-thirties. The important advantages conferred by this form of axle construction are therefore worth enumerating, as follows:

(1) The carrier can be provided with a substantial stiffening rib that completely spans the bearing mountings for the crown wheel and pinion, so that the gears are given maximum support.
(2) In the carrier axle the crown wheel and differential bearings can be widely spaced, which further enhances the stability of the crown wheel.
(3) Maximum lateral support is also provided for the crown wheel and differential bearing cups, which butt against continuous surfaces in the carrier (*Figure 5.4*).
(4) Oil circulation channels can be readily incorporated in the carrier, a feature that is of particular importance for effective lubrication of the pinion bearings, as mentioned earlier.
(5) From a manufacturing point of view, the carrier-type axle can be readily adapted to a wide range of wheel-track widths by alteration in length of the axle tubes and half-shafts, while retaining a common carrier assembly.

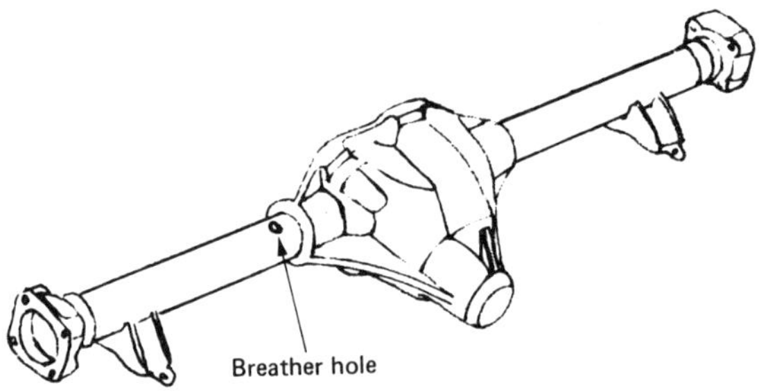

Figure 5.13 Carrier-type axle casing

5.6 AXLE-CASING BREATHER

An appreciable temperature rise accompanies the normal operation of the final drive, this being especially true since the advent of motorways where vehicles can run at sustained high speeds. Axle-oil temperatures can in fact exceed 100°C and it is significant that among the various tests applied by research laboratories to axle lubricants, there is one that evaluates their thermal stability at a controlled temperature of typically 135°C.

Clearly then, some means of venting the axle interior to the atmosphere must be provided, otherwise there would be a danger of pressure building up in the axle and forcing oil past its seals. For a motor-car rear axle this venting may be arranged either simply by a small diameter hole in the axle tube, or by the once-traditional drilled 'breather plug' screwed into the final-drive housing.

On commercial-vehicles a 'breather valve' may be fitted, which is so constructed as to allow heated air to pass out of the axle, but at the same time prevent the ingress of dust and moisture. When the vehicle is operating in extremely dirty conditions, the manufacturer may recommend that the breather valve be removed and thoroughly cleaned at regular intervals.

6 Suspension mechanisms and tyres

6.1 SPRINGS AND SHOCK-DAMPERS

When the road wheels rise and fall over surface irregularities, the springs momentarily act as energy storage devices and thereby greatly reduce the magnitude of loading transmitted by the suspension system to the vehicle structure. Springs that utilise the elastic properties of metal, rubber and gas are variously employed in motor-vehicle suspension systems, the actual choice made being determined largely by versatility in application and best economy of material in terms of energy storage per unit volume. It may be recalled that the metal leaf spring has already been dealt with in § 6.1 of *Vehicle Technology 1*.

6.1.1 Coil springs

Helical compression or 'coil' springs are probably now the most widely used type of suspension spring for motor-cars. In comparison with a leaf spring, the coil spring can store more than twice the amount of energy per unit volume of material and it possesses minimal internal friction. However, the vertical space requirements of a coil spring are greater and its inherently low resistance to buckling is such that it can function as a spring medium only when used in conjunction with a separate wheel-locating linkage.

A coil spring is produced from a length of round wire wound into the form of a helix (*Figure 6.1*) and is installed in a manner that loads it as nearly as practicable in the axial direction. When the spring is compressed, a twisting moment is imposed on the wire as a result of the axial load acting at a distance equal to the mean radius of the coil turns. It may be provided with either two 'flat', one 'flat' and one 'tangent tail', or two 'pigtail' ends. For the second arrangement, the flat end remains stationary relative to the car structure while the tangent tail end seats against a helical-shaped abutment on the moving suspension linkage. This prevents rotational creep of the spring in service, which otherwise can affect the installed rate of the spring. The third arrangement of two pigtail ends is generally used where the ends of the spring are positively clamped to their stationary and moving abutments. Since this form of spring-end also permits a reduction in the fully compressed or 'solid' length of the spring, it finds application where the vertical space to accommodate the suspension spring is strictly limited. As in the case of leaf-spring installations, rubber isolators are usually incorporated in the coil-spring mountings to reduce the transmission of high-frequency noise through the suspension system.

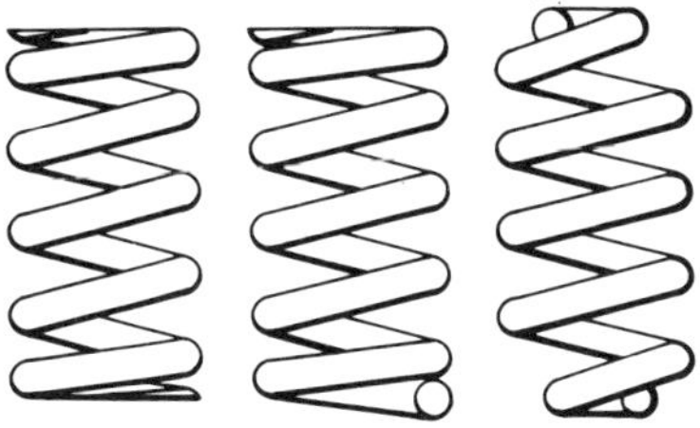

Figure 6.1 Types of suspension coil spring

Several examples of conventional application of the coil spring to IFS systems were illustrated in § 6.4 of *Vehicle Technology 1*. A less common example chosen for this present book is of interest, because it shows a coil spring acting against the upper link of an unequal transverse links IFS system (*Figure 6.2*). An advantage conferred by this 'overhead' mounting of the coil spring is that the suspension loading is distributed over an appreciably larger area of the body front structure, so that the need for a heavy front cross-member is

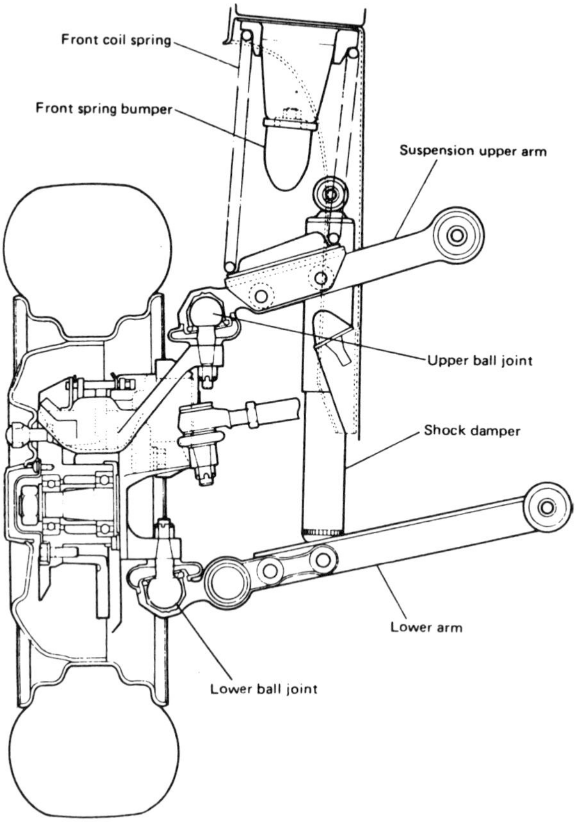

Figure 6.2 Unequal transverse links IFS with overhead coil spring (*Suzuki*)

avoided. Another advantage is that it allows the lower link more freedom for longitudinal compliance, the purpose of which was explained in § 6.4 of *Vehicle Technology 1.*

6.1.2 Torsion-bar springs

This type of spring has not been so widely adopted as the helical or coil spring for motor-vehicle suspension systems, although it is used extensively in a secondary role as an anti-roll bar. In common with the coil spring it possesses minimal internal friction and can only function as a spring medium when used in conjunction with a separate wheel-locating linkage.

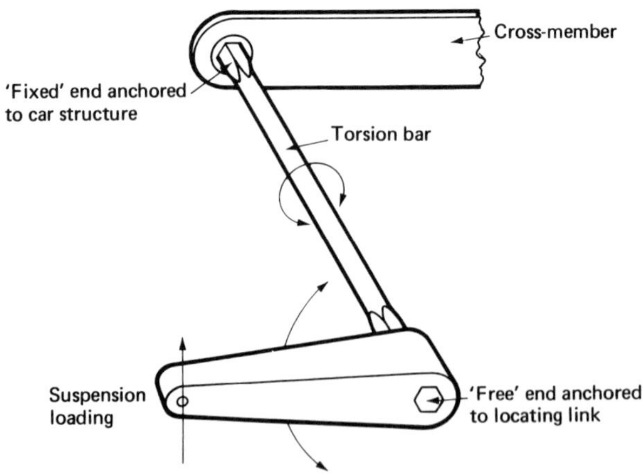

Figure 6.3 Torsion-bar springing in principle

Its method of application is such that one end of the bar is clamped to the car structure to prevent rotation (except for adjustment purposes) and the other end is connected to a lever arm that receives suspension loading. With rising of the road wheels a twisting moment is thus imposed on the bar as it performs its function of energy storage. In this context, the torsion bar offers in the region of a fourfold increase in energy storage per unit volume of material over that of a leaf spring. This further gain in energy storage is mainly attributable to there being less inactive or 'dead' material in the bar-end connections, as compared to that in the end turns of a coil spring.

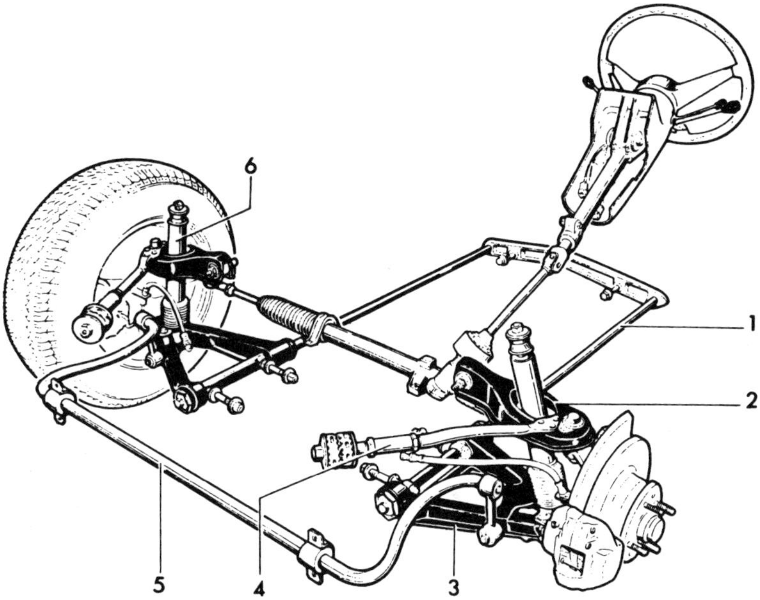

1 Torsion bar
2 Upper wishbone
3 Lower wishbone
4 Castor rod
5 Anti-roll bar
6 Shock-damper

Figure 6.4 Torsion-bar springing in practice (*Alfa Romeo*)

For anchorage purposes the torsion bar is provided with end connections of either serrated or hexagonal form. An advantage of the former is that it requires the least diameter bar ends, while the latter has the production merit of requiring no further machining after grinding of the bar stock and upsetting the ends.

To reduce stress concentration in the critical regions of the end connections, the main body of the bar merges very gradually into the increased diameter ends. Another stress-reducing measure involved in the manufacture of torsion bars results in their being 'handed' side-

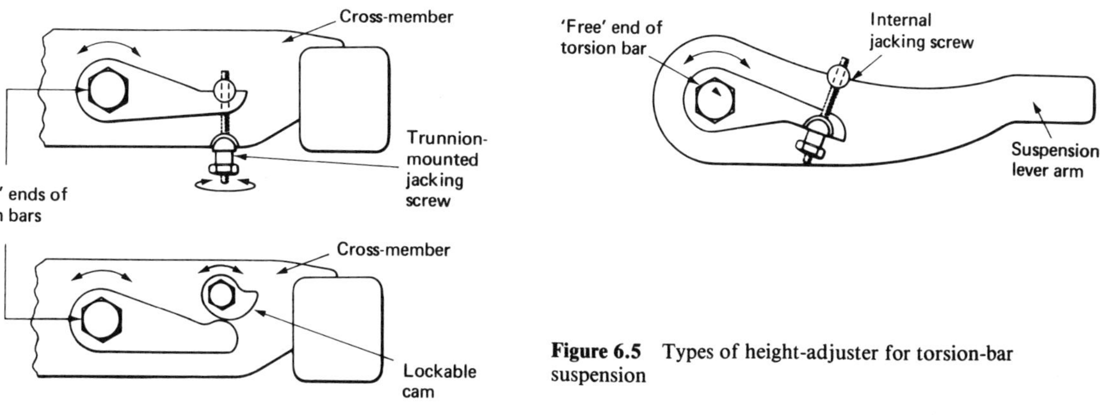

Figure 6.5 Types of height-adjuster for torsion-bar suspension

to-side of the car, so it is essential that a bar is correctly identified (usually by means of colour-coded markings) for whichever side suspension it is intended.

Perhaps it should be mentioned here that to minimise settling of leaf, coil and torsion-bar springs in service, each is subjected during manufacture to an operation known as 'scragging' or 'pre-setting'. This process subjects the spring to repeated overloading of sufficient magnitude to cause yielding in any highly stressed regions. A more uniform distribution of stress is thus obtained upon removal of the load, so that settling is less likely to occur and also the fatigue life of the spring is improved.

For convenience of setting the 'ride height' or 'standing height' of the car, a manual adjustment for winding the bar is provided at either the 'fixed' or, less commonly, the 'free' end. It may take the form of a reaction lever bearing against a trunnion-mounted jacking screw, or alternatively against a lockable eccentric cam. Yet another method is to have a different number of serrations at each end of the bar to provide a vernier setting facility.

In IFS systems the torsion bar is used mainly in conjunction with unequal-length transverse links, where it has the merit of contributing to the basic simplicity of the layout. It is also compatible with front-wheel drive because this type of spring does not get in the way of the drive shaft. A torsion bar is generally claimed to be less bulky than its near relation the coil spring (the material in each is stressed principally in torsion so that a torsion bar may conveniently be regarded as an 'unwound' coil spring), although in most installations space must be found under the body floor to accommodate its length. With longitudinal mounting of the torsion bars there is the advantage that the suspension loads are transferred to the main part of the body structure, which is better able to contain them. On the other hand torsion bars are noise transmitters and it is for this reason they are now usually anchored in a cross-member that is rubber-mounted to the body structure. Another advantage of the torsion-bar spring is that it contributes to a reduction in unsprung mass of the suspension system, the significance of this being mentioned in § 6.5.1 of *Vehicle Technology 1.*

With so many good points, why then is the torsion-bar type of spring not more widely used? The answer to this question is almost certainly that it is more expensive both to manufacture and to build into the car.

6.1.3 Rubber springs

The use of rubber as a spring medium has found relatively limited application in the field of motor vehicle suspension systems, although there have been a few notable exceptions, the best-known of these being the Moulton 'cone spring' applied originally to the then BMC Mini in the late nineteen-fifties. In a secondary role rubber has, of course, been used extensively as a spring material for suspension 'bump' and 'rebound' buffers. These act as progressive limiting stops to prevent metal-to-metal contact between the wheel-locating linkages and the car structure at the extremities of suspension travel.

There are two factors that have tended to mitigate against the widespread application of rubber springs in vehicle suspension systems. One concerns their settling caused by what are termed 'primary creep' and 'secondary creep' of the rubber. The former arises from static loading on the spring and the latter from dynamic

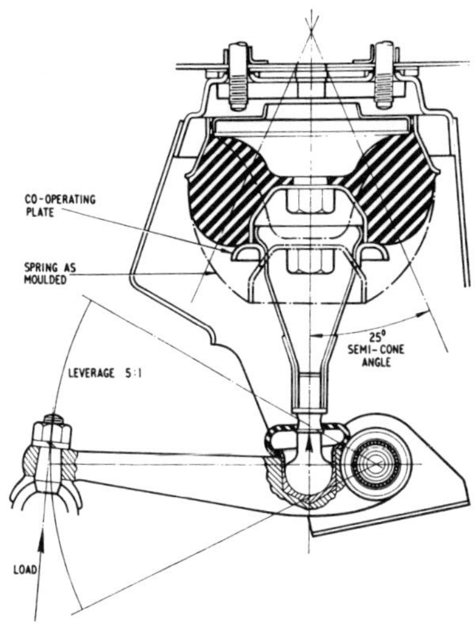

Figure 6.6 Moulton rubber cone spring installation (*BMC* Mini)

loading. Primary creep in a rubber spring is mainly responsible for any reduction in ride height or standing height of the vehicle in service, but since this effect tends to occur early in the life of the spring it can usually be catered for by setting or 'trimming' the vehicle initially high on its suspension. The other factor concerned is perhaps less significant and is the stiffening in action of the spring at low ambient temperatures.

On the credit side, a rubber spring has a very high energy-storage capability and possesses a useful amount of inherent damping due to the internal friction of its material. It also contributes to the reduction of noise transmission through the suspension system.

The familiar Moulton cone spring (*Figure 6.6*) consists of a natural rubber element bonded between inner and outer pressed-steel cones. To provide the required load and deflection characteristics and also to lend inherent stability to the spring, the rubber element assumes a form reminiscent of a 'ring doughnut' although perhaps we should be more formal and describe it as 'frusto-conical'. By virtue of its shape the rubber element is therefore loaded partly in shear and partly in compression, so as to provide the best compromise between greatest energy storage and long service life. The purpose of the 'co-operating plate' is to stiffen the action of the spring under high loading and thereby eliminate the need for a separate bump buffer.

6.1.4 The need for shock-dampers

Shock-dampers play an important role in the motor-vehicle suspension system and work in conjunction with the springs. The need for 'damping' the action of the springs can be readily visualised by considering the gymnastics of a very simple vibratory system consisting of a mass supported by a helical extension spring.

If a disturbing force momentarily displaces the mass from its position of rest, then not only will the spring-restoring force return the mass to its original position, but also the inertia of the mass (or reluctance to change its state of motion) will cause it to overshoot. The mass will then continue to oscillate with diminishing amplitude either side of its original position until the vibration eventually dies out or 'decays'. This must occur, of course, because of the natural damping effects of air friction and of internal friction in the spring material. In fact these frictional effects have 'damped' the energy of vibration by dissipating it in the form of heat energy.

With practical anti-vibration systems found in engineering, and certainly in the case of the motor-vehicle suspension system, it is necessary to reduce the amplitude of vibration far more rapidly than is possible with the 'natural' damping effects just described, otherwise the vehicle ride and road-holding would prove totally unacceptable. Hence, it is usual for some form of 'deliberate' damping to be introduced into the system and this is the duty performed by the familiar shock-dampers. At this point it may be appreciated that the universal description of the shock-damper as a 'shock-absorber' is really a misnomer, since it is the suspension springs and tyres that actually 'absorb' the road shocks and the shock-dampers that 'damp' the oscillations of the springs and their associated masses. However, we must accept that the designation 'shock-absorber' has long since come into general usage.

6.1.5 A modern shock-damper

Shock-damping is performed in the suspension system of the modern motor-vehicle through the medium of 'hydraulic' friction, which is created by the resistance offered to a fluid that is forced by movement of a piston to flow through a submerged orifice.

The 'telescopic' type of shock-damper has been used in the suspension systems of most motor-vehicles designed from the nineteen-fifties onwards. It is usually described as 'direct-acting' by virtue of its operating piston and cylinder being directly attached between the vehicle structure and the suspension linkage. In contrast, an earlier form of hydraulic damper was the 'lever arm' type where the piston and cylinder assembly was almost invariably mounted on the vehicle structure and intercepted movements of the suspension linkage via a lever arm and connecting link. An inherent disadvantage of this type of damper was that any reduction in damper resistance was considerably amplified at the road wheel, because of the intervening lever-arm arrangement.

Returning to the telescopic damper, this is installed either as an individual unit, or forming an integral part of a transverse link-and-strut MacPherson suspension system. It may also be of either the original 'twin-tube' (*Figure 6.7*) or the later 'mono-tube'

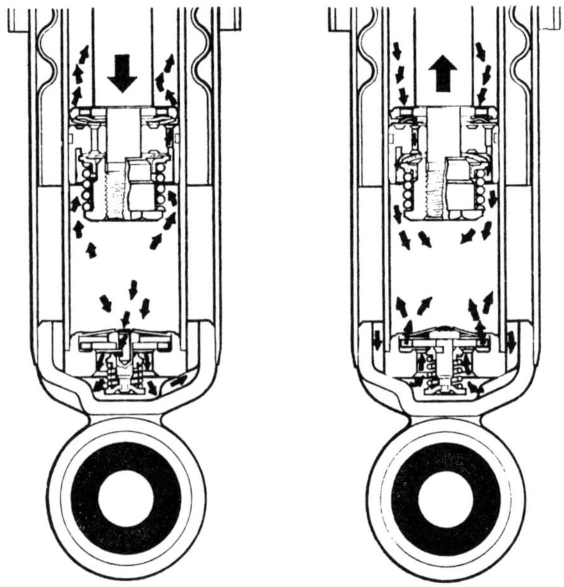

Figure 6.7 Operation of a twin-tube hydraulic shock-damper (*Woodhead*)

construction, the latter being pressurised internally to prevent aeration of the fluid.

At this stage we shall confine our attention to the basic operation of a twin-tube damper, as follows:

Bump movements of the suspension: As the damper retracts, a two-way flow of oil is established through the hydraulic resistances offered by valves in the piston and in the base of the cylinder. That is, the damper fluid passes upwards through the stationary piston and downwards through the base of the rising cylinder, the volume of fluid passing through the latter to the reservoir being equal to that displaced by the piston-rod entering the head of the cylinder. To ensure that the increasing space above the piston is maintained completely full of fluid, the hydraulic resistance in the base of the cylinder must be greater than that in the piston and therefore governs the damper resistance to bump movements of the suspension.

Rebound movements of the suspension: As the damper extends the fluid passes downwards through the piston and upwards through the base of the cylinder, the volume of fluid returning from the reservoir to the cylinder similarly being equal to that of the retracting piston-rod. Conversely, in this case, it is the hydraulic resistance in the piston that must be the greater and therefore governs the damper resistance to rebound movements of the suspension.

In actual practice there is, of course, a good deal more sophistication in the design of a telescopic shock-damper than the foregoing simple description might suggest, especially in respect of valve operation. Indeed it should be recognised that the hydraulic damper is a precision piece of equipment that may have to withstand up to 20 000 oscillations per mile and on bad roads the temperature of its outer casing may reach 100°C or more in dissipating the energy of vibration as heat to the atmosphere.

6.2 INDEPENDENT REAR-SUSPENSION (IRS) SYSTEMS

Whereas a few car manufacturers, and especially some in America, continue to mount the rear wheels on a sprung live axle, many others have long since adopted various forms of independent rear-suspension (usually abbreviated to IRS). It first became widely used by German and Middle European manufacturers during the nineteen-thirties, notably in the designs of Drs Ferdinand Porsche and Hans Ledwinka, but this development did not really gain popularity elsewhere until some 30 years later.

The chief benefits to be expected from using a modern IRS system are generally concerned with the inter-related qualities of ride, handling and, in the case of rear-wheel-drive cars, traction. Ride comfort in particular should benefit from the reduction by about one-half in the unsprung mass of the suspension mechanism, resulting from the final-drive assembly being mounted on the vehicle structure. Also an increase in useful space within the body rear portion is implicit with IRS, since the propeller shaft and final-drive assembly do not rise and fall in sympathy with the suspension movements of the rear wheels.

The improvement in traction to be expected with independently-sprung and driven rear wheels deserves a few words of explanation. In § 4.4 of *Vehicle Technology 1* mention was made of the antics performed by the live axle of a Hotchkiss-drive system during acceleration and braking. Taking matters a little further, we find that during acceleration the axle casing rocks on its springs not only in opposition to crown-wheel torque, but also to a lesser extent (as related to the final-drive gear ratio) in sympathy with pinion torque. In other words, the tendency during acceleration is to press one rear wheel harder against the ground and to lift the other one off it. This effect, combined with the one mentioned previously, can cause the axle and wheels to writhe about a conical path and generate an unstable handling condition known as 'axle tramp'. Although this state of affairs may to some extent be alleviated by additional means of axle control, as later described in § 6.17, such misbehaviour is absent from IRS systems. The reason for this is, of course, that the final drive is divorced from the road-wheel mountings and is attached to the vehicle structure, the drive to the wheels being transmitted through universally-jointed drive shafts.

6.2.1 Types of IRS

At this stage we are required only to identify the various systems by

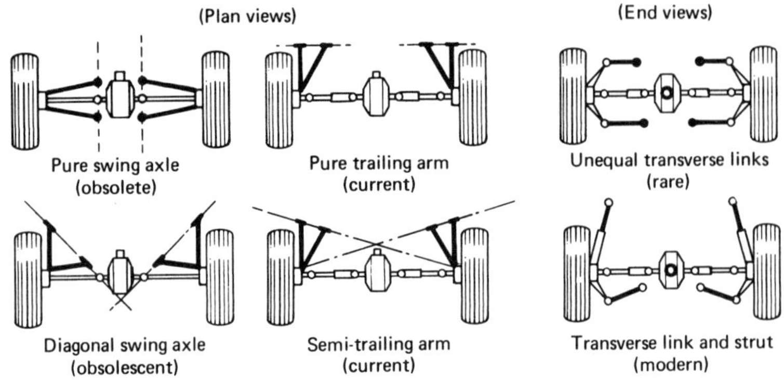

Figure 6.8 Identification of IRS systems by defining their pivot axes

their basic geometric layout, rather than to relate the geometric effects caused by rise and fall of the road wheels to vehicle handling qualities. The systems, illustrated in *Figure 6.8*, may be classified as follows:

(1) Swing axle—pure and diagonal.
(2) Trailing arm—pure and semi.
(3) Unequal transverse links—the shorter of which may be the drive-shaft (e.g., Jaguar).
(4) Transverse link and strut.

The last-mentioned type of IRS, and one which is becoming increasingly wider used for the non-driven rear wheels of FWD cars, is sometimes referred to as a 'Chapman strut', so-named after the Lotus car designer who first applied the MacPherson transverse link-and-strut principle to rear-wheel suspension.

6.3 TYPES OF SPRING USED WITH IRS

The simple pure trailing arm system of IRS, which is also often employed for the relatively lightly laden rear wheels of front-wheel-drive cars, lends itself fairly readily to all types of spring, including in the past a transversely-mounted leaf type. For rear-wheel-drive cars the widely used semi-trailing-arm system of IRS is generally associated with coil springs, similar to when it first appeared on the Renault Fregate model in the mid-nineteen-fifties. Coil springs are also usually employed with the increasingly popular transverse link-

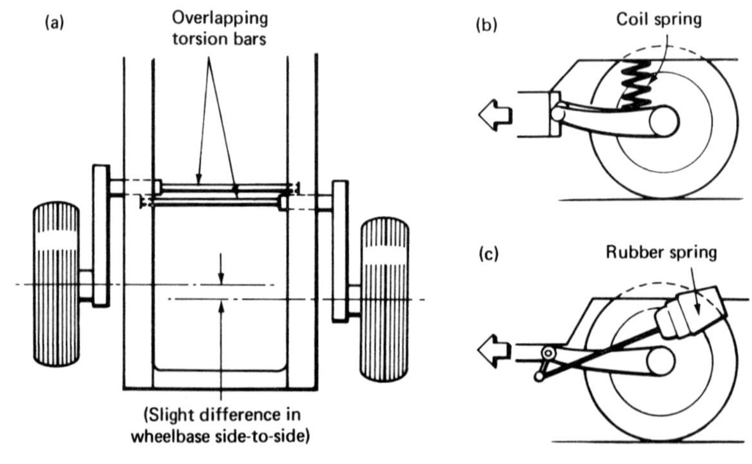

Figure 6.9 IRS with different types of spring. (a) Pure trailing arm with torsion-bar springs (plan view), (b) semi-trailing arm with coil spring (side view), (c) pure trailing arm with rubber springs (side view)

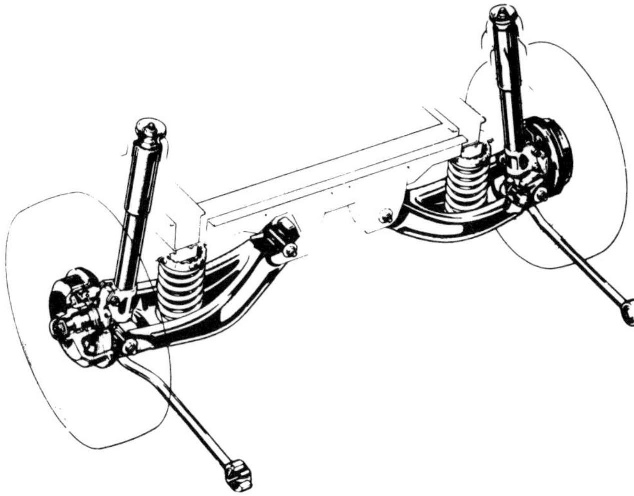

Figure 6.10 A modern strut-type IRS with coil springs (*Ford*)

and-strut system of IRS, although instead of encircling the strut they may act against the lower links to reduce intrusion into rear body space (*Figure 6.10*). This system also lends itself neatly to a transverse-leaf spring installation (*Figure 6.17*).

6.4 FLUID-GAS SUSPENSION SYSTEMS

In considering the operation and construction of this type of suspension system, it is important first to establish the division of labour between the fluid and the gas, as follows:

(1) The compressible gas constitutes a self-contained springing medium.
(2) The incompressible fluid provides the means of transmitting suspension loads to the springing medium.

All gases are readily compressible without ever being damaged and this property has led to the limited use of an inert gas (non-flammable nitrogen) as a spring medium in fluid-gas suspension systems. As compared to other spring media, the gas spring also possesses the highest energy-storage capability. Furthermore, the nature of its application to the fluid-gas suspension unit is such that a 'self-levelling' facility may readily be incorporated in the fluid section. This ensures that the whole range of suspension movement always remains available regardless of vehicle load and was a feature pioneered by the French Citroën Company in the early nineteen-fifties. Alternatively, the fluid sections of the front and rear suspension units on each side of the car may be 'interconnected' in the interests of providing a relatively pitch-free ride, as in the well-known 'Hydragas' system that was introduced in 1973 by Moulton Developments for certain British Leyland cars.

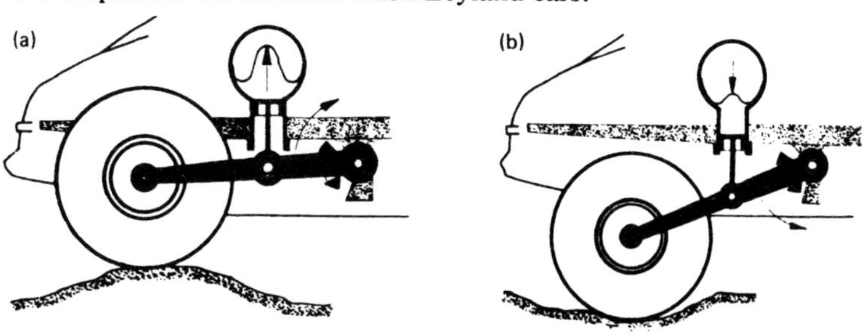

Figure 6.11 Basic action of a fluid-gas suspension system (*Citroën*). (a) Bump, (b) rebound

6.4.1 Examples of fluid-gas suspension units

The Citroën type of hydro-pneumatic strut (*Figure 6.12*) is connected between the car structure and the wheel-locating linkage. At the head of each strut is a spring unit formed in the upper half of a spherical chamber, which is pressure-charged with nitrogen and sealed from the lower half by a flexible diaphragm. Communicating with the other side of the diaphragm is a fluid-filled cylinder, in which a piston reciprocates in sympathy with suspension movements. Since the column of trapped fluid is virtually incompressible, its displacement by the piston further compresses the fixed mass of gas in the spring unit to absorb the road shock taken by the rising wheel. A damper valve is incorporated at the base of the lower half of the spherical chamber, thereby creating a resistance to fluid flowing to and from the spring unit. To provide a self-levelling facility, the column of fluid trapped in the strut is adjusted in length by either releasing, or admitting fluid under pressure from a separate energy source.

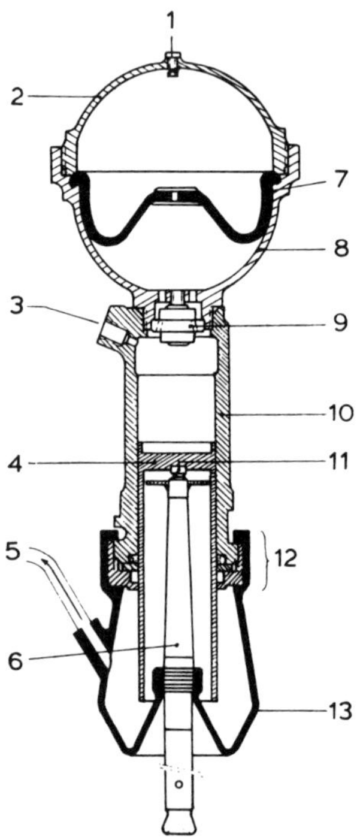

1 Filler screw
2 Upper half-sphere
3 High-pressure fluid inlet
4 Piston
5 Overflow return
6 Piston rod
7 Diaphragm
8 Lower half-sphere
9 Damper
10 Cylinder
11 Thrust pad
12 Sealing system
13 Dust shield

Figure 6.12 Cross-section through a hydro-pneumatic strut (*Citroën*)

The Moulton type of Hydragas unit similarly has at its head a fixed-mass gas spring, but the fluid acting against it is contained within a 'displacer body' in which a tapered plunger reciprocates according to suspension movements. Again, a damper valve is incorporated at the base of the lower half of the gas spring chamber, or 'egg' as it is sometimes called. The sealing function between the displacer body and the tapered plunger is performed by a second diaphragm of the rolling-lobe type, such that the extent to which it envelopes the plunger head determines the effective area acting upon the fluid. As mentioned earlier, the Hydragas units may be interconnected on each side of the car, so as to suppress pitching. If this begins to occur, a restoring force is created by virtue of the difference in effective diaphragm areas and hence fluid displacements

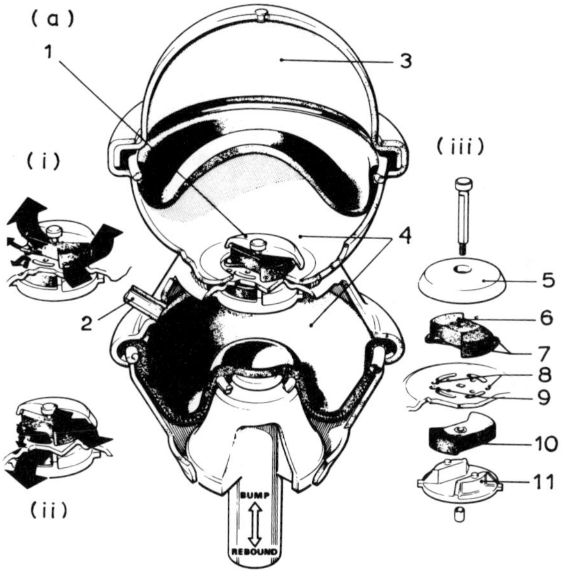

(i) Bump flow through damper
(ii) Rebound flow through damper
(iii) Exploded view of damper valve
1 Damper valve
2 Interconnection pipe
3 Gas
4 Fluid
5 Bump housing
6 Bump compression block
7 Flaps
8 Bump bleed
9 Main bleed
10 Rebound compression block
11 Rebound housing

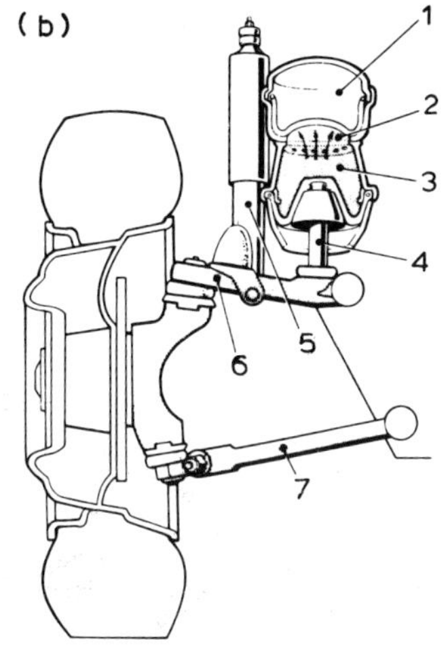

1 Nitrogen spring chamber of Hydragas unit
2 Drilled stainless steel diaphragm in lieu of damper valve
3 Fluid chamber
4 Push-rod
5 External telescopic damper
6 Upper suspension forging
7 Lower suspension forging

Figure 6.13 Comparison between early and recent Hydragas suspension types. (a) Early type (*BL Austin* Allegro), (b) recent type (*BL* Mini Metro)

created by the rising and falling of the tapered plungers, since the excess fluid displaced serves further to compress the gas in both spring units.

6.5 CAMBER, KING-PIN AND CASTOR INCLINATIONS

In order to take into account the various features that constitute the steering or 'front-end geometry' of a vehicle, it is necessary not only to consider the system in plan view, as in § 7.2 of *Vehicle Technology 1*, but also in end and side views since we are here concerned with camber, king-pin and castor inclinations.

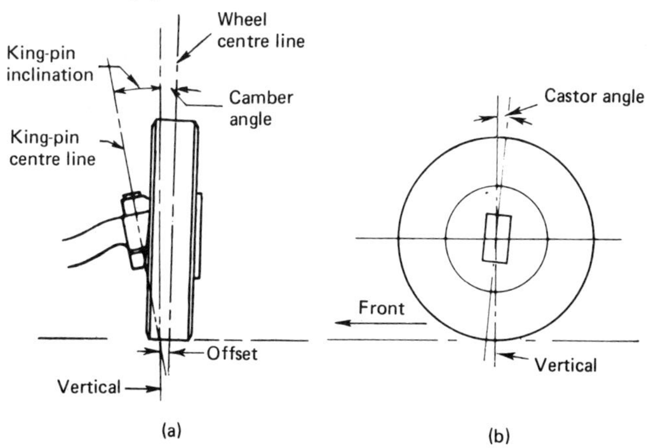

Figure 6.14 Front-end geometry of beam-axle suspension. (a) Camber and king-pin angles, (b) castor angle

6.5.1 Camber angle

This is the inclination of the road-wheel centre plane relative to the vertical, as viewed from the front of the vehicle. The camber angle is considered to be 'positive' if the top of the wheel leans outwards and 'negative' if it leans inwards. Historically, a noticeably positive camber was given to the front wheels of motor cars of an earlier era (as any reader who has ever seen one of the legendary pre-war Bugatti cars head-on will agree!), its purpose being to maintain the wheels normal to the surface of the then steeply cambered or 'crowned' roads.

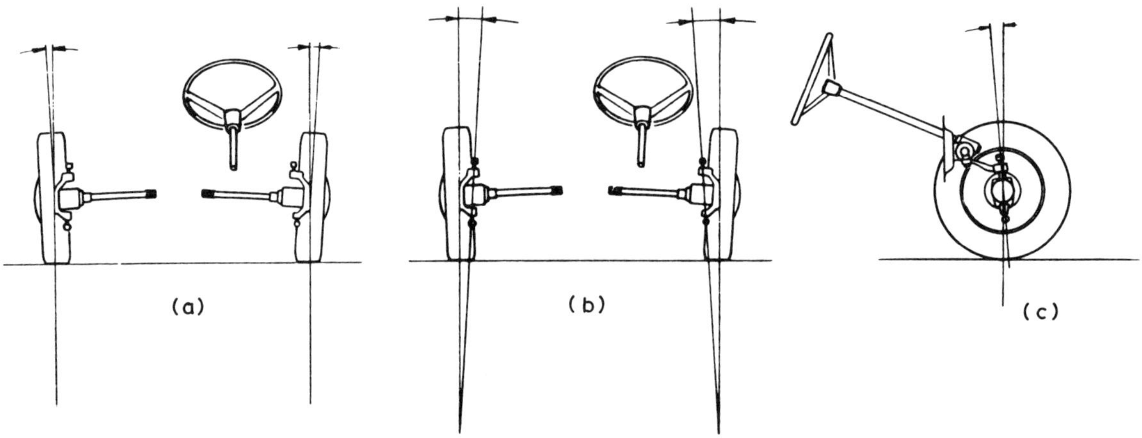

Figure 6.15 Front-end geometry of IFS. (a) Camber, (b) king-pin angle, (c) castor (*SAAB*)

In modern automotive practice, however, the front wheels are usually provided with either a zero, or a small camber angle of up to about 1° positive, at the normal standing height of the vehicle. The tendency of a cambered wheel to roll in a direction towards which it is leaning, thereby producing a side force or 'camber thrust' can have two beneficial effects on the steerable wheels. First, it tends to counteract any small sideways forces imposed on the tyres by ridges in the road, and second, it induces a small lateral pre-load in the steering linkage. The overall effect is therefore to promote stable, straight-ahead running.

6.5.2 King-pin angle

This is the inclination of the road-wheel swivel axis relative to the vertical, as viewed from the front of the vehicle. It should be observed that the steering swivel axis coincides with the centre-line of the king-pin on beam-axle front suspensions (thus the origin of the term 'king-pin angle'), whereas it passes through the centres of the spherical joints on modern IFS systems.

In conventional practice, the angle of the swivel axis is such that it leans inwards at the top. Although this inclination is in any event dictated by structural features, it is also usually desirable from handling considerations. With regard to the former, it is the physical bulk of the brake assembly sandwiched between the road wheel and its pivot arrangements that demands some inclination of the swivel axis. This is because when the swivel axis is projected down to ground level it must either coincide with, or more usually be slightly offset from, the centre of the tyre contact area. Further consideration will be given to this matter at a later stage.

From a handling point of view, an incidental advantage of king-pin inclination is that it confers a self-aligning torque on the wheels, since their sloping swivel axes cause the front of the vehicle to be lifted slightly as they are steered towards either lock. As the vehicle sinks to its normal front standing height, the wheels thus have a natural tendency to return to their straight-ahead position, this being effective in both the forward and reverse directions. A king-pin angle in the range of 6 to 12°, at the normal standing height of the car, is typical of modern practice.

6.5.3 Castor angle

Not to be confused with king-pin angle, this is the inclination of the road-wheel swivel axis relative to the vertical, as viewed from the side of the vehicle. It is considered to be 'positive' if its intersection point with the ground lies ahead of the vertical centre-line through the wheel and to be 'negative' if it lies behind. The effect of a positive castor angle is to create at ground level a moment arm distance or 'mechanical trail', such that the centre of the tyre contact area always tends to trail behind the swivel axis where it meets the ground. Exactly the same principle is involved in the design of the inclined steering head and forks of pedal and motor-cycles. A familiar household application of this principle is, of course, found in the construction of furniture castors.

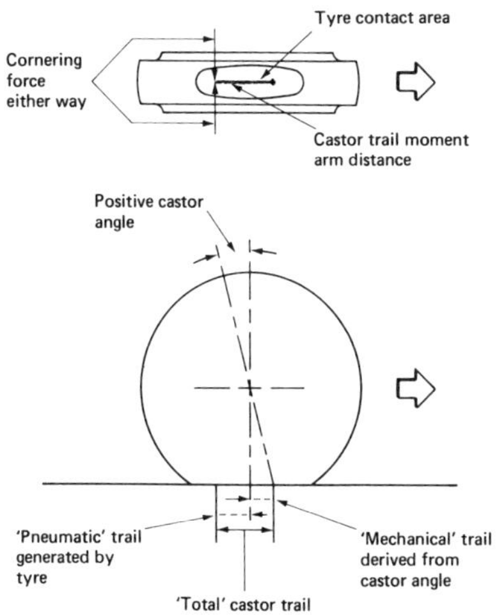

Figure 6.16 The principle of castor action

In the case of motor vehicles, however, the 'mechanical trail' due to castor angle is merely additive to the much greater 'pneumatic trail' generated by the tyre itself, so that in combination the two trails present a moment arm distance or 'castor trail' at which cornering force acts to provide a self-aligning torque on the steered wheel. Castor effect from this source becomes most effective at the relatively small steering angles used during normal and high-speed driving. This is in contrast to the castor effect derived from the previously-mentioned king-pin angle, which is most effective at the relatively large steering angles confined to low-speed driving.

On modern vehicles fitted with radial-ply tyres a positive castor angle in the range of 1 to 3° is typically specified. There are generally two reasons why a castor angle appreciably greater than this may be employed. The first is now the less common one of providing sufficient castor effect for the lightly laden front wheels of rear-engined cars and the second is the rather subtle one of keeping the steered wheels more nearly upright in the presence of body roll. (It may be recalled from § 6.4 of *Vehicle Technology 1* that there is usually some tilting of independently-sprung front wheels with body roll.)

The cambering effect that accompanies high castor angles is, of course, most readily observed during parking manoeuvres and explains the somewhat drunken posture of the wheels as they lean increasingly towards the direction in which they are being steered. A glance at any 2CV Citroën parked with its front wheels turned to either full lock should make this effect abundantly clear!

6.6 IRS WHEEL ALIGNMENT

The correct alignment of independently-sprung rear wheels assumes just as much importance as it does for the front wheels. Here again it must be appreciated that the wheel-locating linkages for IRS systems cannot be made completely rigid, especially where rubber bushings are present at the pivot points. As a result there is a tendency for driven rear wheels to toe-in during acceleration and to toe-out during deceleration and braking. (The non-driven rear wheels of FWD cars tend to toe-out from the effects of rolling resistance and braking.)
If, as in the case of the front wheels, we are going to set the rear

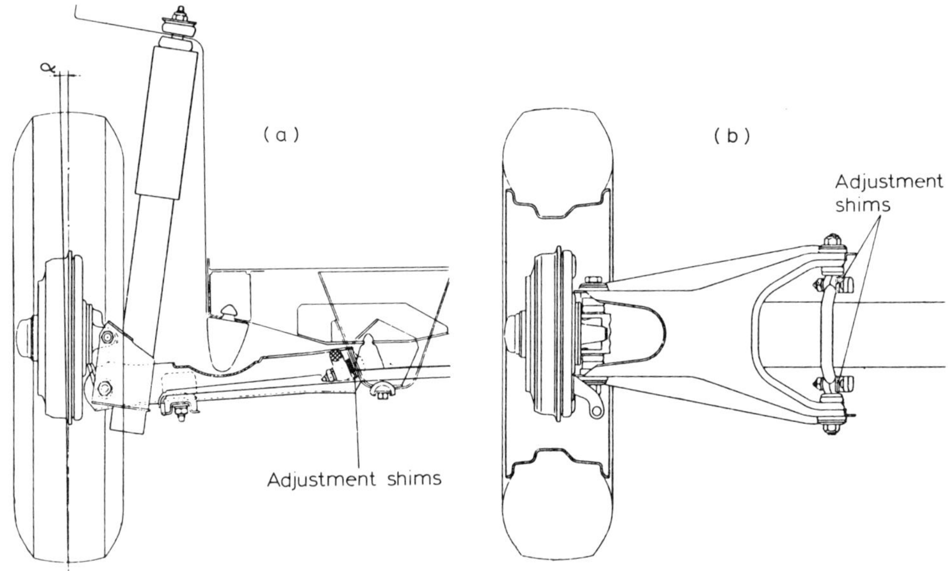

Figure 6.17 Camber and toe adjustment for strut-type IRS. (a) Rear view, (b) plan view (*FIAT*)

wheels deliberately misaligned when stationary, so that they run parallel when the vehicle is in motion, then which condition are we going to compensate for—toe-in during acceleration or toe-out during deceleration? In fact it is the latter condition that must be avoided, since it can provoke an unstable oversteering condition when entering a corner. For this reason, it is therefore usual to find that independently-sprung rear wheels are provided with a small amount of static toe-in.

6.7 ADJUSTING ALIGNMENT OF INDEPENDENTLY-SPRUNG REAR WHEELS

The popular semi-trailing arm type of IRS especially requires accurate toe-setting and is generally provided with some means of adjustment at either, or both, the bearing pivots for each arm. Fore-and-aft repositioning of one of the pivot pins may be accomplished by the use of a slot and a fixed location for its eccentrically-headed

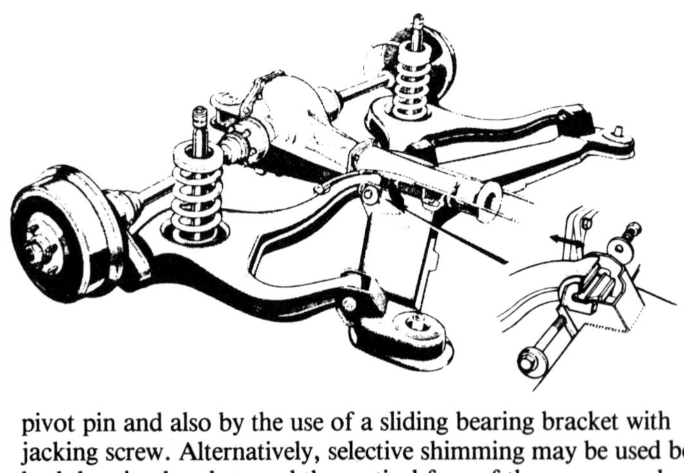

Figure 6.18 An example of toe adjustment for semi-trailing-arm type IRS (*Ford*)

pivot pin and also by the use of a sliding bearing bracket with jacking screw. Alternatively, selective shimming may be used between both bearing brackets and the vertical face of the cross-member to which they are attached. This form of adjustment can also be applied to the bearing brackets of a transverse link-and-strut system where it serves the dual purpose of setting toe and camber of the rear wheels.

6.8 RADIAL-PLY TYRES

The basic constructional features of a radial-ply or 'braced-tread' tyre for motorcar application were described and illustrated in § 6.6.4 of *Vehicle Technology 1*. It may therefore serve a useful purpose here to illustrate, in comparison, the special features of a heavy-duty commercial vehicle tyre of similar construction. (*Figure 6.19*).

6.9 TYRE TREAD PATTERNS

On a wet road surface the water film acts as an excellent lubricant for the tread rubber, so that in the absence of some form of tread pattern to disperse this water the grip or adhesion of tyre to road would be drastically reduced and the safe handling of the vehicle jeopardised. The main purpose of the tread pattern is therefore to provide both a wiping action and a drainage facility between the tyre and road.

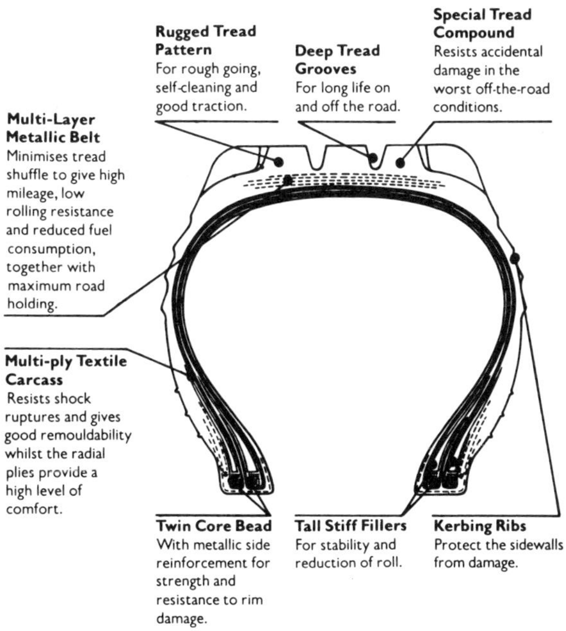

Figure 6.19 Cross-section of a radial-ply truck tyre (*Pirelli*)

A 'flooded', as opposed to a 'wetted', road surface poses a much more serious threat to tyre adhesion. It is then possible for the combined effects of speed and depth of water to be such that a worn tyre is no longer able to remove quickly enough the water from its contact area with the road, especially if the remaining depth of tread pattern is approaching the legal limit stated in § 6.12. In this event, the tyre can become entirely supported by the film of water with consequent loss of control. This condition is known as 'aquaplaning' or 'dynamic hydroplaning', since a water wedge builds up between the tyre and road to generate a hydrodynamic upthrust on the tyre (*Figure 6.20*). Indeed, the action can be likened to that of hydrodynamic lubrication of a plain bearing, as described in § 5.2 of *Engine Technology 1*.

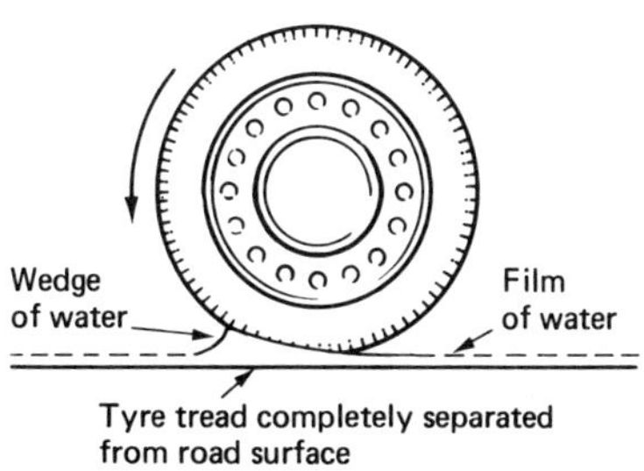

Figure 6.20 Tyre aquaplaning effect

Additional requirements of the tyre tread pattern are that it should provide quiet running and acceptable wear characteristics. A non-

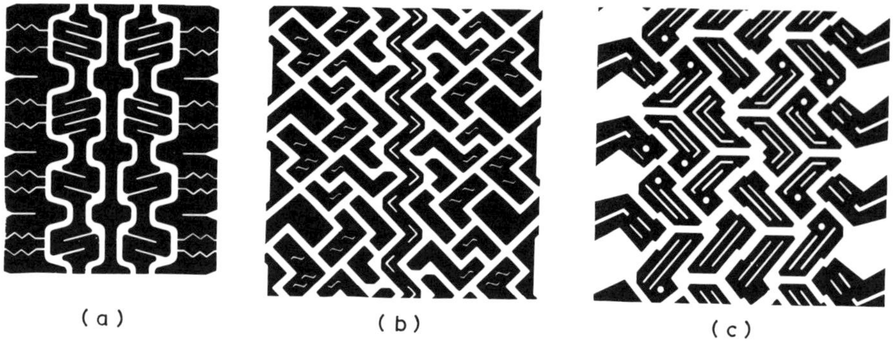

Figure 6.21 Examples of modern tyre-tread patterns. (a) Every-day motoring, (b) high performance, (c) all-seasons mud and snow (*Pirelli*)

linear (not repeating itself at equal intervals) pattern is the most widely used method of avoiding a constant frequency note that could provoke tyre noise. Adequate support for the tread pattern elements is also essential to prevent their excessive distortion, which could otherwise lead to high rates of tread wear.

To achieve adequate traction in adverse off-the-road conditions, such as encountered during rallying, a relatively deep open tread pattern is required. The concentration of loading on the individual tread blocks then causes them to penetrate the material covering the ground to the firmer surface beneath. At the same time, the wide spacing, shape and flexibility of the tread blocks assist in evacuating the penetrated material, which may be deep mud and snow.

Figure 6.22 Computerised research into tread patterns (*Goodyear*)

Tread patterns fall into three basic categories, 'circumferential ribs', 'continuous zig-zag' and 'individual blocks', all of which are usually supplemented by a series of 'sipes' or knife cuts that function as tiny storage reservoirs as they pass through the tyre to wet road contact area. Although it is probably true to say that every conceivable form of tread pattern has, at some time, been evaluated by the various tyre manufacturers, development work still continues on this theme except that nowadays it is aided by sophisticated computer programming (*Figure 6.22*).

6.10 TYRE SPEED MARKINGS

This topic was introduced in § 6.8 of *Vehicle Technology 1*, and here it is presented in an up-dated tabulated form for ease of reference (*Figure 6.23*).

NEW MARKINGS (Service description)	Driving speed limit		CURRENT MARKINGS (Size designation)		
Speed symbol	mph	km/h		Diagonal	Radial
L	75	120	— — M + S*	(10″ rim) (10″ rim)	
M	80	130			
N	85	140	— — M + S*	(12″) (12″)	
P	95	150	S — — M + S*	(10″) (13″ and above) (13″ and above)	SR reinf. M + S*
Q	100	160	S	(12″)	SR M + S*
R	105	170			SR reinf.
S	112	180	H S	(10″) (13″ and above)	SR
T	120	190	H	(12″)	HR M + S*
U	125	200	H	(13″ and above)	
H	130	210			HR
	Over 130	210	V	(13″ and above)	VR

* M & S or M-S or MS (winter-type tyres) also permitted

Figure 6.23 Table of tyre-speed markings (*Goodyear*)

6.11 TYRE MATERIALS

These may be classified under the three headings of 'rubber compounds', 'cord materials' and 'bead wire', and it is worth reflecting that although their various properties differ considerably they must work in harmonious physical contact throughout the life of the tyre.

Rubber compounds: These refer to the composition of the 'rubber mix' used in the construction of a tyre. At this point it is first necessary to understand what we mean by the term 'rubber'. A useful engineering definition attributable to the American Society for Testing and Materials tells us that rubber is 'a natural or synthetic material that can be or is already vulcanised to a state in which it has high extensibility and forcible, quick retraction'. For many years it was the generally-held opinion of tyre manufacturers that the use of natural rubber was essential for reasons of reliability, but the increase in speed and cornering capabilities of high-performance cars designed towards the end of the nineteen-fifties brought about the serious development of synthetic rubber materials for tyres. Their eventual acceptance made possible safer braking and cornering on wet roads, a reduction in squeal when braking or cornering on dry roads, and an increase in tyre life, although natural rubber still plays an important role in tyre construction.

The principal synthetic rubbers or 'polymers' used in the construction of modern tyres are typically those known to the tyre engineer as 'Styrene-butadiene rubber' (SBR) and 'Polybutadiene' (PBD), their relative proportions determining the compromise between tyre grip and wear resistance. These polymers, combined

with other chemical agents such as carbon blacks for abrasion-resistance and oils for flexibility, produce the desired properties of the compound.
Cord materials: These serve as reinforcing components in the carcass plies and tread-bracing layers of a tyre. The carcass unites the tread and the beads and is produced from rubberised cord plies that impart the necessary strength to the tyre while also being highly flexible. Each reinforcing cord consists of twisted together yarns of either rayon, nylon or polyester fibre, these materials posessing in varying degree superior physical properties to the cotton originally used for this purpose. For current versions of radial-ply tyre, the virtually inextensible stabiliser belt typically consists of two layers each of rubberised high-tensile steel and nylon cords.
Bead wire: A rubberised high-tensile steel wire, larger in diameter than that used for tread-bracing purposes, is used to wind either round, square or hexagonal 'bead bundles'. These are necessarily the least flexible component of the tyre structure, since there must be no relative movement between the inflated tyre and wheel rim.

6.12 THE LAW AND VEHICLE TYRES

The following constitute illegal practices with respect to the use of vehicle tyres:

Types: If the tyre is unsuitable for the use to which the vehicle is being put or is incompatible with the types of tyres fitted to the other wheels, as previously explained in § 6.7 of *Vehicle Technology 1.*
Pressure: If the tyre is not sufficiently inflated to make it fit for the use to which the vehicle is being put.
Tread: If the tyre does not have a tread depth of at least 1 mm across a continuous band of at least three-quarters of the breadth of the tread and round the entire outer circumference, or, where the original tread pattern only covered three-quarters of the breadth of tread, does not have a depth of at least 1 mm from the base of the grooves.
Structure: If the tyre has a break in its fabric or has a cut that exceeds 25 mm or 10% of the section width, whichever is the greater, measured on the outside of the tyre and deep enough to reach the body cords. Also if the tyre has any portion of its ply or cord structure exposed, and further, if it has any lump or bulge caused by separation or partial failure of its structure.

6.13 LOAD-CARRYING CAPACITY OF A TYRE

This may be considered in terms of its size and ply rating, as follows:

Size: The choice of tyre size must primarily take into account the weight per axle of the vehicle. In practice, it is the 'economic load' that is used as the basis for tyre selection. This term is applied to the load capacity beyond which a tyre of larger size or with a stronger carcass will be more economical under medium service conditions.
Ply rating: This indicates the strength of a tyre and is a term used to identify a given tyre with its recommended load and inflation pressure when used in a specific service. At one time the ply rating directly related to the number of cord plies used, but the advent of cord ply materials stronger than the original cotton meant that a tyre with fewer plies could have the same load-carrying capacity. Car-tyre equipment is usually of four-ply rating with two actual carcass plies being used in the typical modern radial-ply tyre.

6.14 REASONS FOR BALANCING ROAD WHEELS

Until the early nineteen-fifties very little was heard about 'wheel balancing', although it was not unknown to those concerned with the servicing of a relatively select few high-performance cars. Since then, of course, the need for balancing the wheel and tyre assemblies has become a fact of life for all concerned. Historically, the reasons for this are basically twofold:

(1) It was found that the more softly sprung car with IFS displayed a greater sensitivity to any imbalance of the wheel and tyre than was previously the case with a relatively stiffly sprung beam-axle suspension system.
(2) The ill effects of any such imbalance became more pronounced with the trend towards smaller diameter and hence faster rotating wheels, which from tyre-loading considerations were also increased in width.

Technically, there are two types of wheel imbalance and these are known as 'static' and 'dynamic'. (See *Figure 6.24.*)

6.14.1 Static imbalance in principle

It is a common experience for a schoolboy to invert his cycle and then proceed to hand-pedal it until the back wheel rotates so fast that the rear end of the machine begins to jump up and down of its own accord. This is caused, of course, by the out-of-balance forces generated by the revolving weight of the tyre valve and provides a simple example of 'static' imbalance of a wheel and tyre assembly. The reason why this particular condition of imbalance should be called 'static' can be explained as follows.

If the schoolboy transferred his attentions to the more freely running front wheel, he would find that from whatever stationary position the wheel is released it would slowly turn until the unbalanced valve portion finally settled at bottom dead centre. Fairly obviously this state of affairs could be corrected by attaching a counter-balance weight on the wheel rim, equal to that of the tyre valve, directly opposite to the latter.

Summing-up, a condition of static balance is achieved when there is an equal distribution of weight about the axis of rotation of the wheel and tyre.

6.14.2 Dynamic imbalance in principle

Returning once again to the schoolboy hand-pedalling his inverted cycle, if for some peculiar reason the tyre valve emerged not from the centre of the rim, but was offset to either side of it, then the out-of-balance forces generated would not only cause the rear end of the machine to jump up and down, but also to shake from side to side due to 'dynamic' imbalance.

To cure both these conditions of imbalance it would be necessary to attach a counter-balance weight, equal to that of the tyre valve, directly opposite and offset in the same direction to the latter on the wheel rim. A state of both static and dynamic balance would then be achieved.

On the other hand, if the counter-balance weight was placed centrally on the rim or offset in the opposite direction to the tyre valve, then static balance would still be obtained but not dynamic and in fact this would be made worse.

Summing-up, a condition of dynamic balance is achieved when there is an equal distribution of weight about the centre plane of the wheel and tyre.

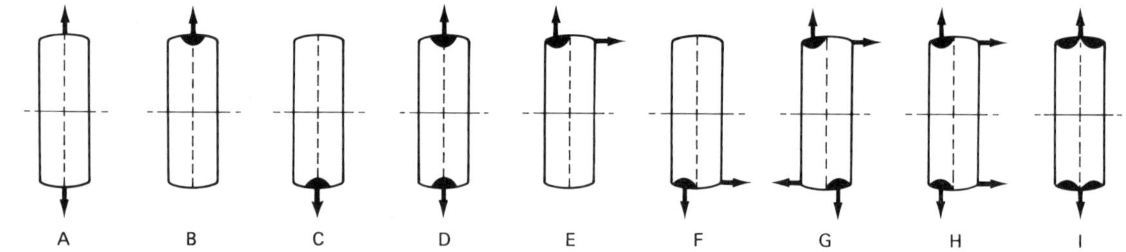

A — 'Ideal' dynamic and static balance
B/C — Static imbalance
D — Static balance restored
E/F — Dynamic and static imbalance
G — Dynamic imbalance made worse but static balance restored
H — Dynamic and static balance restored
I — 'Ideal' restoration of dynamic and static balance

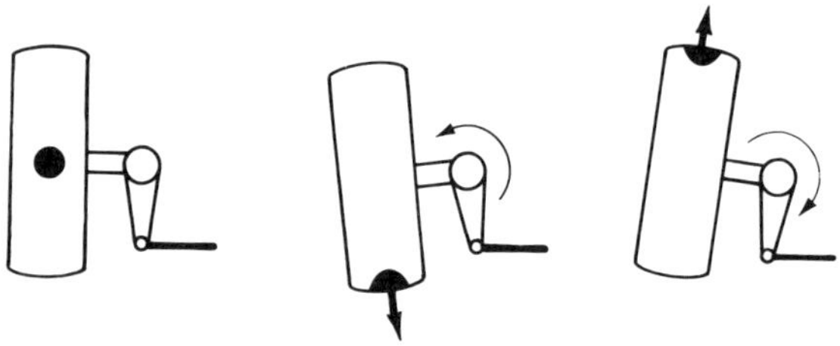

Figure 6.24 Conditions of wheel and tyre balance and imbalance. The smaller diagram shows in plan form how static imbalance can cause wheel shimmy and steering joggles

6.14.3 Wheel imbalance on the vehicle

A glance at any heap of scrapped tyres will reveal that the varied nature of tyre wear is usually such that the distribution of the rubber mass becomes irregular, so as a tyre wears it is likely that the initial balancing of the wheel and tyre by the manufacturer will be disturbed. Static imbalance alone can make itself evident in two ways, as follows:

(1) The vertical out-of-balance forces generated can provoke the wheel into bouncing up and down or 'pattering', which is neither conducive to good road-holding nor to long tyre life.
(2) The corresponding horizontal out-of-balance forces can promote wheel 'shimmy', because they tend to steer the road wheel first in one direction and then in the other about its offset swivel axis. This is signalled to the driver by 'joggles' felt at the steering wheel.

Dynamic imbalance again has the effect of promoting wheel shimmy, except that in this case it is the transverse out-of-balance forces generated that tend to steer the road wheel about its swivel axis.

6.15 WHEEL-BALANCING EQUIPMENT

Modern proprietary wheel-balancing equipment falls into two basic categories, these being known as 'off-the-car' and 'on-the-car' balancers (*Figures 6.25, 6.26*).

As their designation would suggest, off-the-car balancers require that the wheel and tyre assembly be removed from the vehicle for balancing. Their operating principle is typically such that the wheel to be balanced is mounted on a motor-driven horizontal shaft. The bearing housing for the shaft is spring-mounted in the horizontal direction only against a pair of electrical transducers, which produce voltages proportional to the forces imposed upon them by any imbalance of the spinning wheel. This information is interpreted by a small computer so that after the wheel has been braked to a stop, a meter read-out can be taken of any out-of-balance at the inner and outer rims and the appropriate correction weights attached thereto. It is, of course, necessary in the first instance for the operator to program-in the rim position, width and diameter of the wheel.

1 Cubicle
2 Compression springs (mounting)
3 Leaf springs (mounting)
4 Plastic cylinder
5 Position of shaft indicator
6 Shaft drive pulley
7 Transducers
8 Bed casting
9 Ball bearings
10 Mounting shaft for wheel

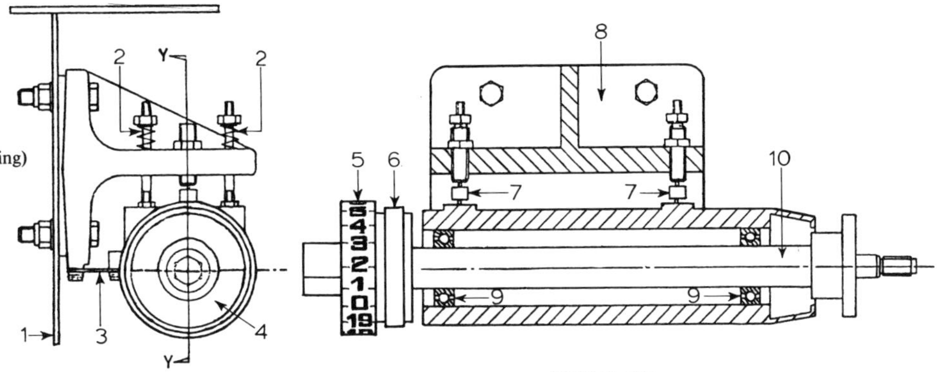

Figure 6.25 Wheel mounting shaft and bearing assembly of an 'off-the-car' balancer (*Churchill*)

In contrast, the on-the-car balancer checks static and dynamic out-of-balance with the wheel on the vehicle. Hence, the hub and brake drum or disc, as well as the wheel and tyre assembly, are compensated for imbalance. Naturally, this condition of overall balance can be lost in the event, say, of the wheel being removed and then not replaced in exactly the same rotational position relative to the hub. With this type of equipment the wheel to be balanced is jacked clear of the ground and, unless it is normally driven through the vehicle transmission system, it is rotated by a mobile, motor-driven, 'wheel spinner' held in contact with the tyre. For sensing any out-of-balance

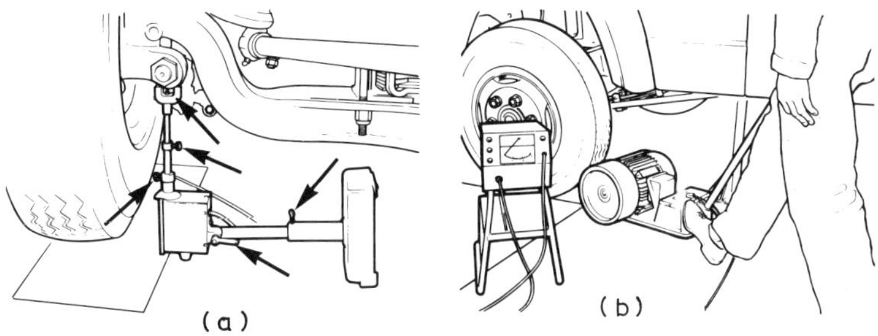

Figure 6.26 An 'on-the-vehicle' wheel balancer in operation. (a) Adjusting the pick-up, (b) spinning the wheel (*Churchill*)

forces acting on the wheel assembly, a transducer 'pick-up' device is attached magnetically beneath the suspension linkage and as near as practicable to the spinning wheel for 'static' balancing, and against the forward part of the brake assembly back-plate or splash shield for 'dynamic' balancing. The equipment is completed by an amplifier, stroboscopic flash unit (similar to an engine-timing light), weight-indicating meter and controls. Any vibration set up by imbalance of the spinning wheel is therefore signalled by the pick-up to the strobe flash and the weight-indicating meter. The maximum unbalanced force shown on the meter and the position on the wheel at which a previously affixed piece of white tape appears 'frozen' by the flashing strobe light are noted. Reference is then made to an 'adjustment of weights' chart (provided with the equipment) before attaching correction weights to the wheel rims.

For all types of wheel-balancing equipment it is vitally important that the particular manufacturer's operating instructions are consulted.

6.16 HUB ASSEMBLIES FOR INDEPENDENT SUSPENSION

In modern practice these may be considered under the following headings:

(1) IFS with non-driven wheels.
(2) IFS with driven wheels.
(3) IRS with driven wheels.
(4) IRS with non-driven wheels.

IFS with non-driven wheels: The hub bearing arrangements for this system (*Figure 6.27(a)*) generally continue to follow those of previous

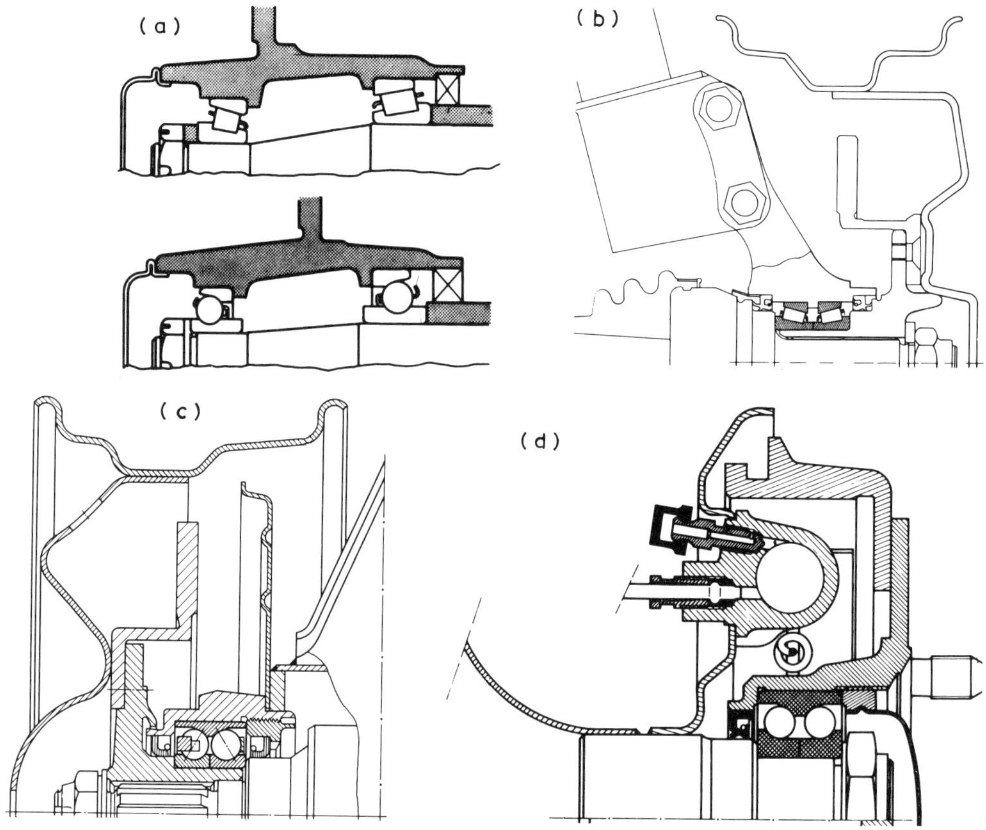

Figure 6.27 Typical hub-bearing arrangements. (a) IFS/non-driven wheels (*RHP*), (b) IFS/driven wheels (*Timken, Ford* Fiesta), (c) IRS/driven wheels (*SKF*), (d) IRS/non-driven wheels (*Citroën*)

beam-axle-mounted wheels. That is, each hub rotates about a 'dead' spindle through the medium of an opposed pair of either angular-contact ball, or tapered-roller, bearings. However, the requirements in respect of their load-carrying capacity have tended to increase, because of the more forward mounting of the engine permitted by IFS. It is usual for the line of action of the wheel load to pass closer to the inner bearing than to the outer, so that the latter mainly performs a stabilising function and can be of smaller size and therefore less expensive. To allow for expansion caused by heat dissipated from the brake assemblies, a very small amount of end-float is usually permitted between the bearings when they are adjusted.

IFS with driven wheels: The development of the front-wheel drive principle where each wheel is secured to a 'live' spindle, which rotates within a pair of bearings mounted in a hub carrier, early created a demand for bearing arrangements that occupied less width than those previously described (*Figure 6.27(b)*). This led to the application of very compact double-row, angular-contact, ball bearing units in which a split inner ring is used to obviate the need for filling slots for the balls. Hence, the optimum number of fairly large diameter balls may be used to increase radial and axial load-carrying capacities, so that a much reduced axial space is required and the wheel spindle can be made short and rigid. Tapered roller bearings may be similarly adapted to give high load-carrying capacity in the minimum space. These modern types of 'hub bearing unit' are set for correct axial clearance during manufacture and require no adjustment in service.

IRS with driven wheels: For driven rear wheels independently sprung by the popular semi-trailing arm system, the modern approach is to use a double-row, angular-contact, ball bearing hub unit similar to that described for IFS with driven wheels (*Figure 6.27(c)*). Since the bearing arrangements in both cases are basically the same, one manufacturer for example uses a common hub bearing unit for the driven front wheels of one model and for the driven rear wheels of another. For heavy-duty applications, the earlier practice of using an opposed pair of either deep groove ball, or tapered-roller, bearings may still be followed.

IRS with non-driven wheels: The lightly-laden rear wheels of FWD cars readily lend themselves to the use of a hub bearing arrangement either similar to that described for non-driven front wheels, or again to a modern double-row, angular-contact, ball-bearing unit (*Figure 6.27(d)*). In the latter case it follows, of course, that the bearing unit is housed within the wheel hub and the split inner rings are mounted directly on the 'dead' spindle.

6.16.1 Lubrication and sealing of hub bearings

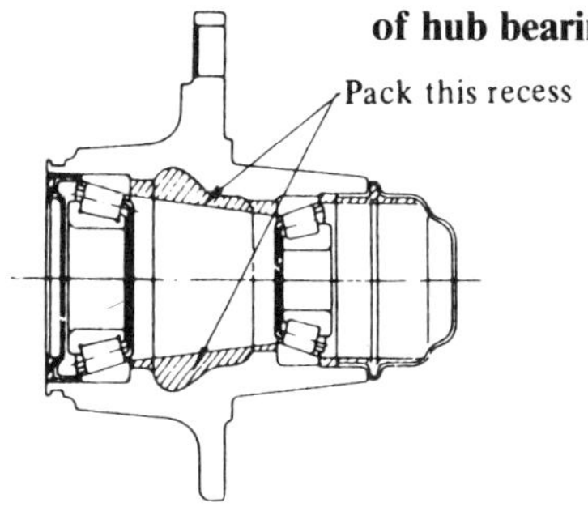

Figure 6.28 Lubrication and sealing of front-hub bearing *(Nissan)*

Hub bearings are lubricated with a multi-purpose lithium-based grease, the characteristics of which combine high temperature and water resistance, together with extreme pressure- and corrosion-inhibiting properties. It is important that the particular manufacturer's recommendations should always be adopted regarding the assembly greasing of hub parts and the extent to which the hub cavity should be charged with grease. The modern 'hub bearing units' described earlier are lubricated for life. Modern sealing practice for hub bearing arrangements of all types is generally to use some form of radial-lip seal, in which the flexible sealing lip is spring-loaded to maintain a light rubbing contact with the surface to be sealed. In 'dead' spindle applications, a hub cap is used to prevent the direct access of dirt and water to the bearing assembly.

6.17 REAR-AXLE LOCATION

Various measures may be adopted to alleviate the inherent 'tramping' tendencies of a live rear axle mentioned in § 6.2. Where still used in modern practice, the live axle usually forms part of a coil spring rear suspension system, a trend that began on American cars as long ago as the late nineteen-thirties. To relieve the coil springs of all axle locating duties, the axle is mounted to the car through a system of pivoted links. These do, of course, exercise a more accurate control over suspension movements of the rear axle than is possible with purely leaf-spring installations. Comprehensive locating systems for live axles may be conveniently classified by their number of longitudinally-disposed (or nearly so) links, of which their may be either two, three or four. At this stage these systems will be identified by their basic geometric layouts, rather than relating their geometric effects to vehicle handling qualities.

Two-link system (*Figure 6.29(a)*): This was an early development and comprised two trailing arms pivoting at their front ends to the car chassis frame and at their rear ends to the axle casing extremities. Rubber bushings were used at both ends of each arm, but those at the rear were so arranged that the axle was restrained from rocking about the arms under torque reaction. Since the two trailing arms could not provide for lateral location of the axle, this duty was performed by what has long been known as a 'Panhard rod'. This consists simply of a rubber-bushed transverse control rod pivoting at one end from the car chassis frame or structure and at its other end to the far side of the axle assembly. It thus confers positive lateral location on the rear axle, albeit at the expense of slight relative sideways movement at the rear end of the car incurred by the arcuate swing of the rod.

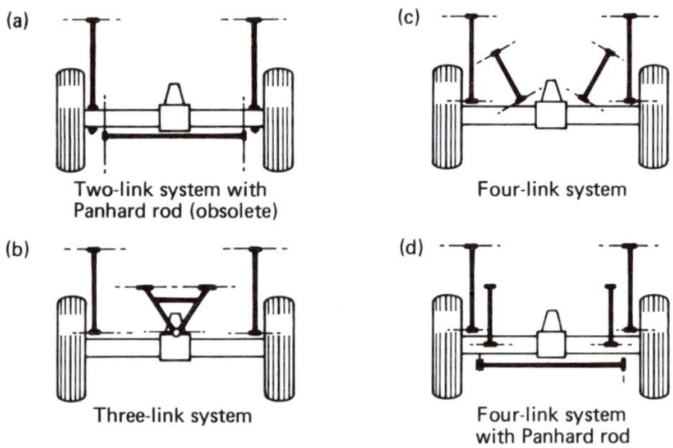

Figure 6.29 Identification of live-axle location systems by defining their pivot axes (plan views)

Three-link system (*Figure 6.29(b)*): This was a later system and again involved the use of two trailing arms, but in this case they are freely pivoted to the axle casing and their action is complemented by a third and shorter trailing arm, such that in combination they form a parallelogram linkage as viewed from the side of the car. The triangulated third arm is usually placed uppermost with a pair of widely-spaced rubber bushes at its forward end. It is ball-jointed at its apex to the centre part of the axle casing, thereby furnishing the necessary lateral location for the axle. Less commonly, the third arm may take the form of a simple connecting link, in which case a Panhard rod must be added for lateral location.

Four-link system (*Figure 6.29(c),(d)*): With the current four-link system, one pair of trailing arms is pivoted below the axle and another pair of either trailing or semi-trailing arms is pivoted above it. If in plan view both upper and lower pairs of arms lie substantially parallel to the centre-line of the car, then a Panhard rod must be added to locate the axle laterally. Alternatively, the upper pair of arms may be disposed in a semi-trailing fashion, so that by virtue of their diagonal mounting they stabilise the axle in the lateral direction. A quieter ride with reduced harshness has often been attributed to the modern four-link system of axle location, this being derived from

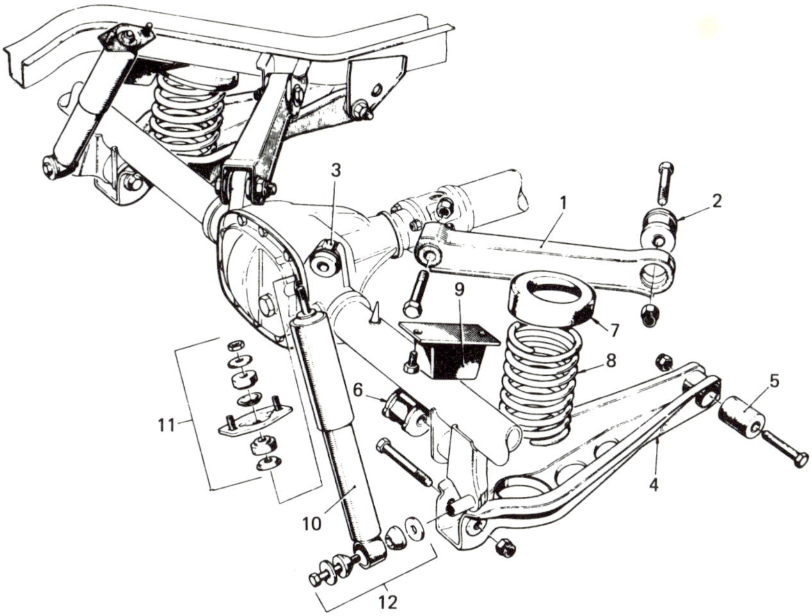

Figure 6.30 A modern four-link rear-axle location system (*Talbot*)

the isolating effect of the multiplicity of rubber-bushed pivot connections involved.

In conclusion, it may be added that the previously-described two- and three-link locating systems are also applicable to 'dead' axle rear suspensions, which are sometimes used to mount the trailing wheels of FWD cars. Also, the four-link system has been used in conjunction with the now rare De Dion rear suspension referred to in § 4.4.1.

6.18 SUSPENSION ANTI-ROLL BARS

An anti-roll bar is basically a torsion bar type of spring with angled ends, and its action is shown in *Figure 6.31.* It is mounted transversely and in such a manner as to connect the locating linkages

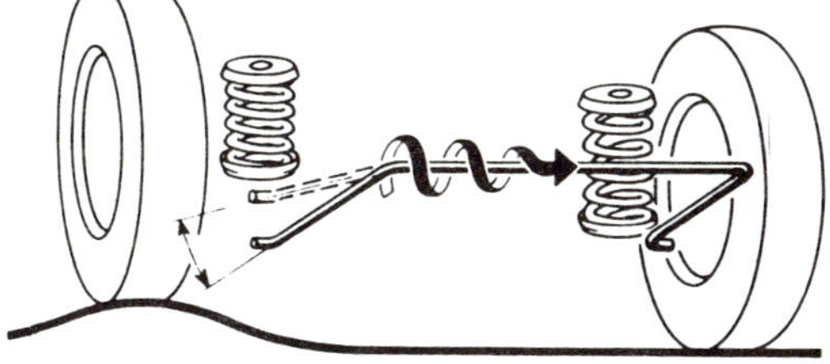

Figure 6.31 Action of anti-roll bar (*Alfa Romeo*)

of opposite wheels, while remaining torsionally free with respect to the car structure. It thus performs the function of resisting rolling, as its name would suggest, but not bouncing motions of the car. However, the amount of roll restraint contributed by an anti-roll bar must necessarily be limited, because it also has the effect of stiffening the suspension in response to single wheel bumps and can therefore compromise ride comfort.

The anti-roll bar was once generally confined in application to the front suspension, because it had the effect of redistributing the overall lateral weight transfer between the front and rear pairs of tyres in favour of promoting understeer. It may be recalled that the influence of tyre slip-angles on understeer was introduced in § 6.7 of *Vehicle Technology 1.* Suffice it to say at this stage that the slip-angle of a tyre can also be increased if the vertical load becomes greater, which is the effect that the addition of an anti-roll bar has on the already heavily-loaded outer wheel of a pair during cornering. In more recent years there has been a trend to include an anti-roll bar in the rear suspension system as well, partly to correct any condition of excessive understeer and partly to reduce roll further.

7 Steering mechanism

7.1 WORM-AND-FOLLOWER STEERING GEAR

At the heart of all motor vehicle steering systems is, of course, the steering gear itself. This comprises some form of reduction gearing to provide the driver with a mechanical advantage in respect of steering the road wheels. To the gear designer such a mechanism would be classified as a 'control gear', since it must have the minimum possible free-play or backlash. This is because its primary job is that of accurately transmitting motion, rather than conveying power. The general requirements for a motor-vehicle steering gear may be summarised as follows:

(1) It should provide for accurate and light steering control without involving excessive movements of the steering wheel.
(2) The transmission of road wheel disturbances to the steering wheel should be minimised, while still allowing the driver to retain an acceptable 'feel' of the varying road surface conditions.
(3) It should not hinder the self-centring action of the steered wheels as the car returns to a straight path.
(4) During parking manoeuvres the driver effort required to steer the wheels should be reduced to a minimum, because it is then that their resistance to being steered is greatest.

Translated into practical terms, these requirements mean basically that the steering gear should possess a high mechanical efficiency in the 'forwards' direction (steering wheel turning the road wheels), but preferably a lower mechanical efficiency in the 'reverse' direction (road wheels attempting to turn the steering wheel). At this stage we shall confine our attention to manual, as distinct from power-assisted, steering gears that utilise some form of 'worm-and-follower' reduction gearing, since the alternative 'rack-and-pinion' type was dealt with in § 7.6 of *Vehicle Technology 1.*

7.1.1 Historical background

The designers of early motor vehicles soon recognised that the steering gear should possess a certain degree of irreversibility or, as previously stated, a lower mechanical efficiency in the reverse direction. In other words they sought to reduce the transmission of road wheel disturbances to the steering wheel. This particular problem could assume quite large proportions, because of the poor state of road surfaces and the limited knowledge then available to relate steering and suspension characteristics to best advantage. It was countered by the general adoption of steering gears that incorporated a simple worm-and-follower reduction gearing of either the 'worm-and-wheel' (or part-wheel known as a 'sector'), or the 'worm-and-nut' variety (*Figure 7.1*). In both designs the worm may be regarded as acting in the manner of a screw thread against its follower. Their reverse efficiency was governed by the angle of the worm threads, the steering becoming less reversible with smaller thread angles. It should be appreciated that total irreversibility of the

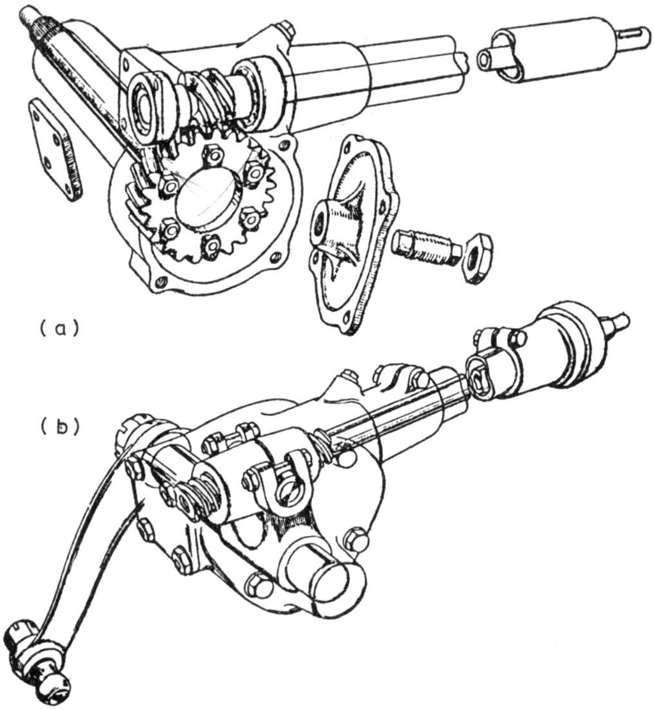

Figure 7.1 Comparison between early worm-and-wheel (a), and worm-and-nut (b), steering gears

steering was not desired since all self-centring ability would then be lost, and in any event the steering gear would have to absorb greater shock loading.

The mid-nineteen-twenties saw the advent of increased section, lower pressure tyres for greater ride comfort. This was further improved by the mid-nineteen-thrities with the gradual adoption of independent front wheel suspension systems, which also contributed to better passenger accommodation by virtue of permitting a more forwards mounting of the engine. In combination, these changes meant that a greater proportion of the car total weight was being carried by the steered wheels on tyres with a substantially increased contact area with the road.

It therefore became increasingly more difficult to keep the driver steering effort within reasonable bounds, especially during parking, without drastically increasing the number of steering wheel turns required from lock to lock. That is, the steering had to be made 'lower geared', which in automotive terminology implies a higher numerical ratio for the reduction gearing. This in turn led to the development, originally in America, of various proprietary 'high-efficiency' steering gears in which rolling, as opposed to sliding, motion was provided between the worm and its follower.

These improved mechanisms reduced steering effort without resorting to excessively low gearing, since forwards efficiency was raised by about 50%. The accompanying increase in reverse efficiency now assumed less importance by virtue of there being better road surfaces and better matching of steering and suspension geometries. High-efficiency worm-and-follower steering gears fall into three basic categories, as follows:

(1) Worm and peg.
(2) Worm and roller.
(3) Worm and recirculating ball-nut.

7.1.2 Worm and peg

This particular type of mechanism was originally devised by a prolific American inventor, David Ross of the Ross Gear & Tool Company, who introduced it in 1923. It possessed the twin virtues of being an unusually compact and robust form of steering gear.

The worm-and-peg steering gear basically comprises a worm with a coarse pitch thread, which intercepts a conical-headed peg carried by a side-mounted lever arm on the drop arm rocker shaft. (Descriptions were given of the manner in which the drop arm of worm-and-follower steering gears can be connected into the steering layouts of beam axle and independent suspension systems in §§ 7.1.1 and 7.1.2 respectively of *Vehicle Technology 1*.) Strictly speaking, this type of steering gear did not become a true high-efficiency mechanism until Ross also mounted the peg in rolling bearings.

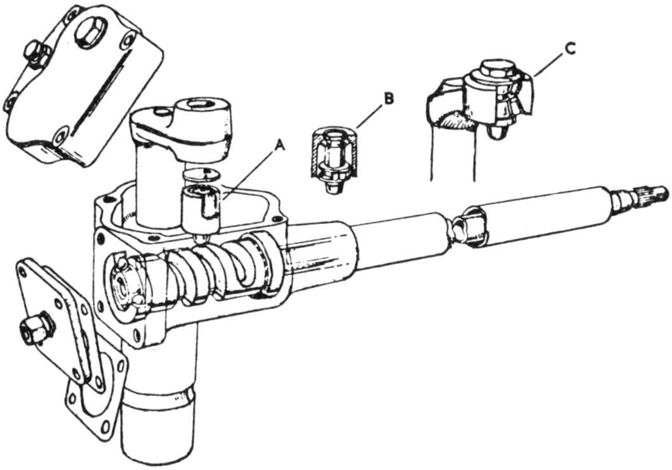

Figure 7.2 Worm-and-peg steering gear. Three basic types of peg assemblies are shown at A, B and C

Although its mode of operation is one of extreme simplicity, in that rotation of the steering wheel compels the peg to follow the helical thread groove of the worm and thus convert the motion of its line of contact into partial rotation of the rocker shaft, its internal geometry is a little more complex. This is because the arcuate movement of the peg causes it to rise and fall relative to the rotational axis of the worm, so that there is an accelerated movement between the worm and the rocker shaft as the steering is turned towards each lock. In driver terms this reducing 'angular ratio' means that the steering is less direct acting about the straight-ahead position, which is the range of movement most used during normal driving, than it is when manoeuvring on either lock. Also the steering will require more effort towards the locks.

To compensate for this effect, the worm thread may be generated with a variable helix angle, which is greatest towards the middle portion of the worm. In this event, the worm can no longer be regarded as a true 'gear' but rather as a form of 'cam', which accounts for this type of steering gear also being known as a 'cam-and-lever' mechanism. Further modifications to the worm-and-peg steering gear that not only maintained more nearly constant the effective leverage of the side lever arm about the rocker shaft, but also permitted a substantial increase in angular capacity (drop-arm travel), were the introduction in the mid-nineteen-thirties of twin and, at a later date, triple pegs.

However, the worm-and-peg steering gear is obsolete for motor-car practice having disappeared from American vehicles by the mid-nineteen-fifties and from vehicles elsewhere by the late nineteen-sixties. The reason for this was the increasing popularity of worm-and-recirculating ball-nut and rack-and-pinion steering gears respectively.

7.1.3 Worm and roller

This type of mechanism was invented by another American, Henry Marles, who in 1919 was the first to substitute purely rolling for sliding contact between the meshing elements of a worm-and-follower steering gear. It evolved through several versions before

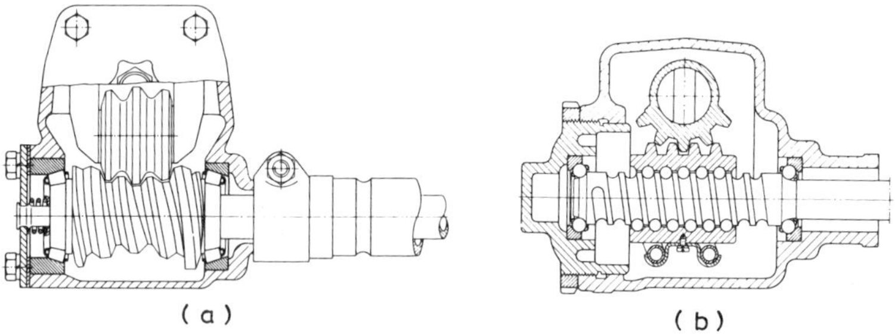

Figure 7.3 Cross-sectional comparison between typical worm-and-roller (a), and worm-and-recirculating ball-nut (b), steering gears (*SKF*)

assuming its most popular form, which comprises an 'hourglass'-shaped worm that engages with a double-toothed roller follower carried between the jaws of the rocker shaft for the drop arm. The periphery of the roller is formed into the contour of a worm-gear tooth.

Its mode of operation is such that around the straight-ahead position of the steering the worm thread bears on the inner flanks of the double-toothed roller, but as either lock is approached it begins to bear on the appropriate outer flank of the roller. Since with an hourglass worm the helix angle of the threads reduces as the pitch diameter increases, the internal leverage of this type of steering gear

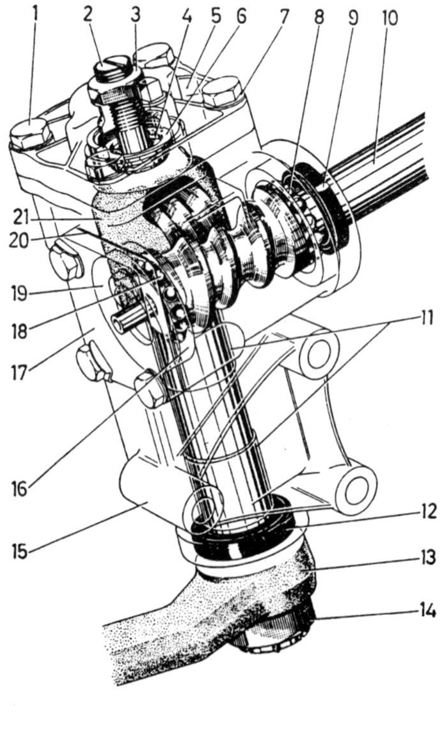

1 Bolt
2 Adjusting screw
3 Locknut
4 Drop arm shaft bush
5 Cover
6 Circlip
7 Tab washer
8 Steering column bearing, upper
9 Steering column oil seal
10 Steering column
11 Drop arm shaft bush
12 Drop arm shaft seal
13 Drop arm
14 Nut
15 Steering box body
16 Steering column bearing shell, lower
17 Steering column cover
18 Steering column bearing, lower
19 Washer
20 Shims
21 Drop arm shaft with roller

Figure 7.4 Identifying the parts of a worm-and-roller steering gear (*Volvo*)

remains substantially constant from lock to lock. A triple-toothed roller is sometimes used to increase the angular capacity of the worm-and-roller steering gear. Similar to the case of the worm-and-peg mechanism, the specialised geometry of the hourglass worm really qualified it as a form of cam rather than a gear, which accounts for it also being known as a 'cam-and-roller' steering gear.

The worm-and-roller type of steering gear continues to find limited application in European and Russian-built cars, but in America it was entirely superseded towards the late nineteen-sixties by the worm-and-recirculating ball-nut mechanism.

7.1.4 Worm and recirculating ball-nut

Introduced in 1939 by the Saginaw Steering Gear Division of the General Motors Corporation, this further American development in high-efficiency steering gears represented a return to the basic worm-and-nut principle, but with the important difference that a chain of ball bearings was interposed between the thread grooves of the rotating worm and its translating nut. Since in this type of construction the ball bearings cannot roll indefinitely around the helical path of the worm, a diagonal return passage must be provided that connects externally the beginning and end of the thread groove in the nut. Hence, the chain of ball bearings rolls within a closed circuit in the nut and replaces sliding by rolling friction against the worm. Either a single, or two independent, ball-bearing circuits may be used according to the severity of duty imposed.

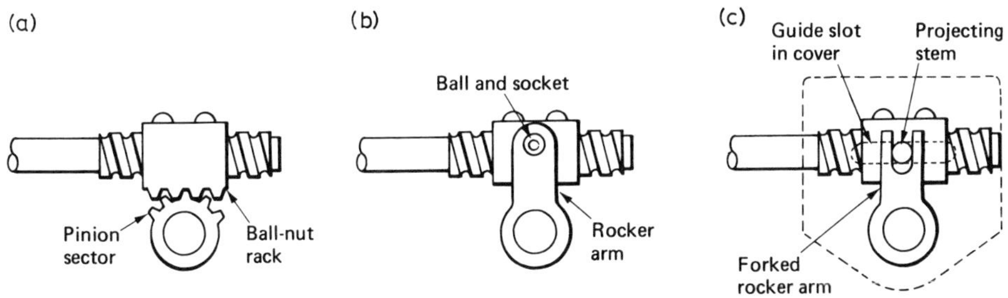

Figure 7.5 Worm-and-recirculating ball-nut mounting systems. (a) Rack mounting, (b) rocker mounting, (c) fork mounting

A not readily apparent, but nevertheless very significant, design detail that contributes to the high-efficiency of the modern worm-and-recirculating ball-nut concerns the profile of their thread grooves. These are typically of 'Gothic-arch' cross-section, so-called after the pointed arch style of architecture that was prevalent in Western Europe from the 12th to 16th centuries. This configuration of non-circular grooves minimises free-play or backlash by reducing the radial and angular freedom of the assembly. It also provides clearance space for any small particles of foreign matter that might enter the thread grooves. Differences in the design of present-day steering gears of this type are mainly concerned with the choice of ball-nut mounting, which may take the form of either a rack, a rocker, or a fork connection, as follows:

Rack connection: This is widely used and featured in the original Saginaw arrangement, where one side of the ball-nut is provided with a toothed rack and meshes with a pinion sector, integral with the rocker shaft for the drop arm. By virtue of this 'rack-and-pinion' type of connection, the angular ratio between the movements of the worm and the rocker shaft remains constant throughout the wide operating range of this type of steering gear (although this characteristic can, in fact, be modified to match the addition of power-assistance). A disadvantage of the rack mounting for the ball-nut is that it tends to produce a less compact steering box than the alternative rocker-and-fork mountings. However, this particular form of steering gear has long since found wide application on American and Japanese cars.

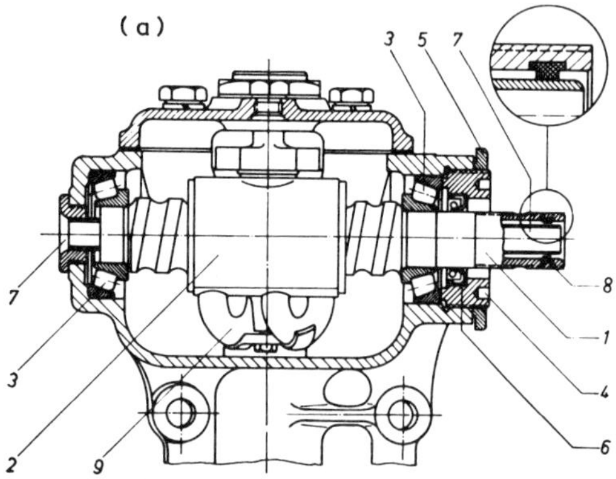

1 Steering worm
2 Steering nut
3 Tapered roller bearing
4 Adjusting ring
5 Hex. nut for adjusting ring
6 Seal ring in adjusting ring
7 Cable tube
8 Seal ring for cable tube
9 Ball guide tube

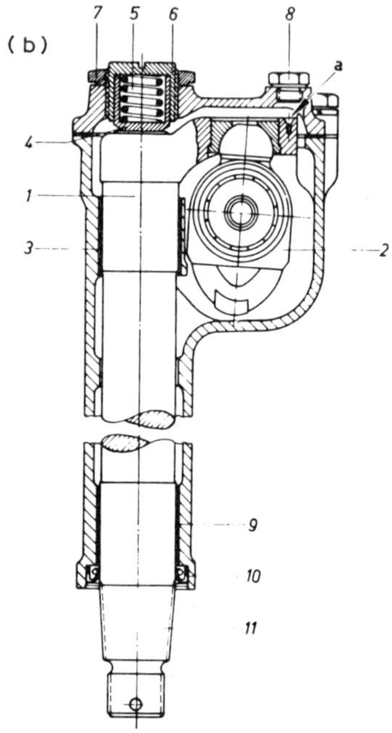

a Centring bore for the centre position of the steering shaft arm
1 Steering shaft
2 Steering nut
3 Bearing bushing, top
4 Pressure sleeve
5 Pressure spring
6 Pressure screw
7 Hex. nut
8 Screw plug
9 Bearing bushing, bottom
10 Seal ring
11 Steering shaft

Figure 7.6 Identifying the parts of a worm-and-recirculating ball-nut steering gear (*Mercedes-Benz*)

Rocker connection: This is effected through a spherical joint and side arm on the rocker shaft. The resulting arcuate travel of the joint is reflected in small angular movements of the nut about its own axis. Hence, there is a reducing angular ratio between the movements of the worm and the rocker shaft as the steering is turned away from the straight-ahead position. In this respect it is similar to the worm-and-single-peg steering gear, except that the reduction in angular ratio differs slightly according to which lock is being approached, because in one direction the angular movement of the nut occurs in the same sense as worm shaft rotation, whereas in the other direction it occurs in the opposite sense.

Fork connection: In this case the ball-nut is provided with a projecting stem upon which is mounted a roller. This is constrained to run in a straight slot machined in the steering box cover plate and thus prevents the ball-nut from turning as it translates. At the base of the roller stem is a chamfered collar, which engages the open forked end of the rocker-shaft side arm. As the steering is turned away from the straight-ahead position, the chamfered collar slides along the fork, thereby increasing the effective length of the side arm. In other words, there is an increasing angular ratio between the movements of the worm and the rocker shaft, which is the opposite characteristic to that found in the rocker mounting for the ball-nut.

7.2 ADJUSTMENT OF WORM-AND-FOLLOWER STEERING GEARS

Two main adjustments are normally incorporated in the modern steering gear, their purpose being to remove all end-play from the worm bearings and to establish correct mesh between the worm and its follower. Another related adjustment concerns the steering 'lock-stops', which are intended not only to limit the maximum angularity of the steered wheels, but also to prevent the steering gear mechanism from over-travelling on extreme locks. The vehicle manufacturer's recommendations must always be strictly followed, of course, when making these important adjustments to the steering.

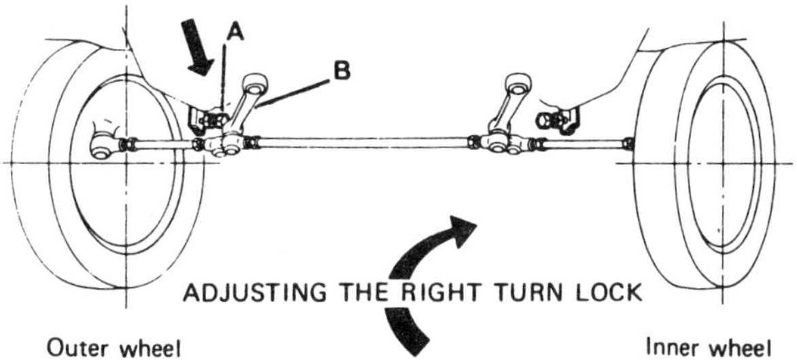

Figure 7.7 Steering lock stop adjusters used with IFS. A — lock-stop adjuster, B — steering arm (*Alfa Romeo*)

7.2.1 Adjustment of worm bearings

This is always the first adjustment to be checked and if necessary performed. The worm is supported between either angular-contact ball, or tapered-roller, bearings mounted face-to-face. These bearings are adjusted to give zero end-float or, in some cases, a slight pre-load. Their adjustment is usually effected at the abutment for the lower bearing outer race, either by means of selective shimming against the end-cover of the steering box or, less commonly, through the medium of a screwed end-plug and locking nut. The steering gear must, of course, be relieved of all external load by removing the connecting linkage from the drop arm. Following adjustment there should typically be a barely perceptible drag as the steering wheel is turned with no more than a thumb and forefinger grip on its rim.

7.2.2 Adjustment of worm-and-follower mesh

Often this takes the form of a single adjuster that not only establishes the correct meshing condition between the worm and its follower, but also controls the end-float of the rocker shaft, which is usually carried in either plain journal bushings or needle-roller bearings.

An important point concerning mesh adjustment is that the follower is generally intended to be in closer contact with the worm around the straight-ahead position of the steering, than it is towards either lock. The purpose of this is to compensate for any greater wear occurring over that range of travel where steering movements are most concentrated during normal driving. It may be recalled from § 7.6.2 of *Vehicle Technology 1* that a similar consideration applies to rack-and-pinion steering gears.

Worm and peg: Any lost motion that develops in the straight-ahead position of the steering can be eliminated by altering the depth of engagement of the peg, since a clearance always exists between its truncated end and the bottom of the thread groove in the worm. For mesh adjustment it was therefore found expedient to allow the side lever arm end of the rocker shaft to bear either directly against a selectively shimmed side-cover of the steering box, or against the end of an adjuster screw carried by the side cover and secured with a locking nut. Hence, the peg can be brought into closer engagement with the worm by the removal of shims or by screwing in the threaded adjuster.

Worm and roller: It has been found expedient in some designs to provide a single adjustment that not only controls rocker-shaft end-float, but also establishes the correct meshing condition between the worm and roller in the straight-ahead position of the steering. The meshing facility is made possible by the intentional offsetting and raising of the roller axis relative to that of the worm, as viewed endwise. To effect the necessary adjustment, a co-axially-mounted threaded stud with thrust collar engages a T-slot in the roller-end of the rocker shaft. Screwing in the threaded stud thus brings the roller into closer engagement with the worm thread groove, the adjustment being retained by a locking nut. In some earlier designs of worm-and-roller steering gear, notably those of American origin, the roller was located centrally below the worm. Here the adjustment took the form of an eccentric sleeve or 'lantern' for mounting the worm-shaft bearings, so that partial rotation of the sleeve in the appropriate direction brought the worm into closer contact with the roller. This form of adjustment was also once used by Rolls-Royce in their version of the Marles worm-and-roller steering gear.

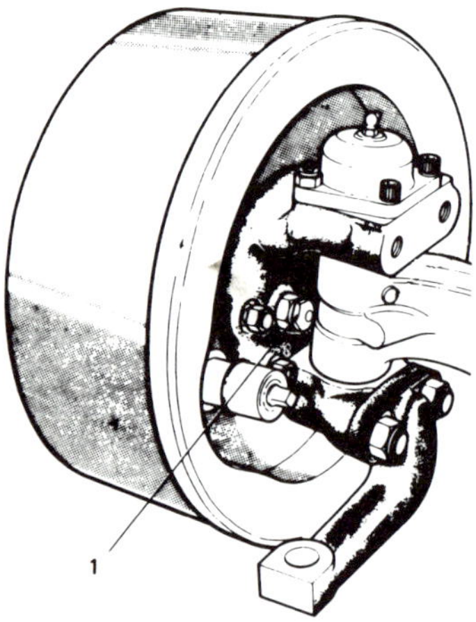

Figure 7.8 Steering lock stop adjusters used with beam front axle. 1 — axle stop adjusting screw (*Seddon Atkinson*)

Worm and recirculating ball-nut: All three types of ball-nut mounting may be brought into closer radial contact with the worm through endwise adjustment of the rocker shaft. For both rocker and fork ball-nut mountings this adjustment can be simply effected by the usual screw and locknut arrangement on the side-cover of the steering box. In the case of the rack ball-nut mounting, the endwise adjustment of the rocker shaft is accomplished by a co-axially-mounted threaded stud, as described previously for the worm-and-roller mechanism. Screwing in the threaded stud has the effect of bringing the teeth of the sector into closer mesh with those of the ball-nut rack (and therefore the ball-nut into closer radial contact with the worm) because the sector teeth are provided with a small wedge angle specially for this purpose. Whichever type of mesh adjustment is provided it is, of course, carried out with the steering in the straight-ahead position (*Figure 7.6(b)*).

A final point in connection with the rocker-mounted ball-nut is that this particular arrangement readily lends itself to a self-adjustment facility. This may be obtained by incorporating a compression spring and plunger within a hollow threaded adjuster, so that once the endwise adjustment of the rocker shaft has been set, the spring will exert a pre-load on the ball-nut and its worm and thus maintain the steering gear free of backlash throughout its full working range (*Figure 7.6 (b)*).

7.3 STEERING AND SUSPENSION BALL-JOINTS

The need for incorporating ball-joints in the steering linkage of both beam-axle and independent suspension systems was explained in § 7.1 of *Vehicle Technology 1.* Before looking at the constructional details of these joints, it may be useful to summarise what is expected of them, as follows:

(1) Accommodate both angular displacement of the steering links and rotational movement of the steering levers, as related to the suspension and steering motions of the road wheels.
(2) Provide freedom from lost-motion or backlash in the steering linkage, so as to maintain accurate steering contol over an extended service life.
(3) Exert a controlled degree of friction damping on the rotational movement of the steering levers, thereby

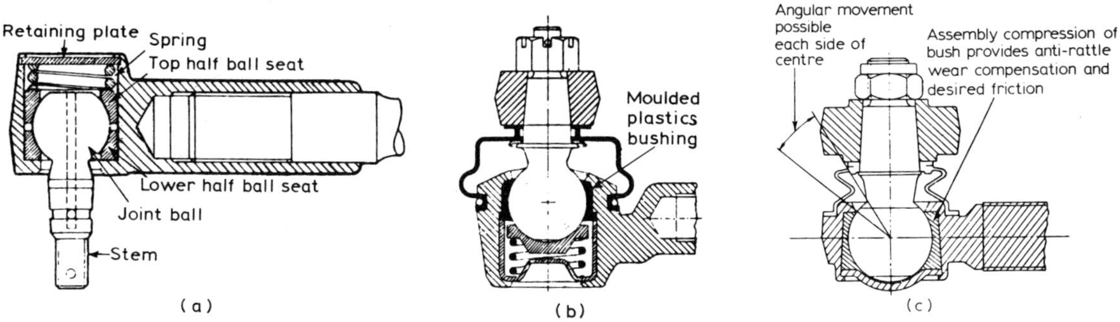

Figure 7.9 The development of steering ball-joints. (a) Early spring-loaded type with metal seatings, (b) later spring-loaded type with plastics seating, (c) modern type with compressed plastics bushing (*Cam Gears*)

suppressing any tendency towards low-speed wobble of the road wheels.

(4) Either eliminate or minimise the frequency of periodic lubrication by virtue of their bearing material properties and sealing capability.

There are basically two forms of steering ball-joint, which may be classified by the direction in which the steering forces are transmitted through the ball-pin, these being known as 'transverse' and 'axial' joints.

7.3.1 Transverse ball-joint

This type of ball-joint is overhung from the steering levers, so that the ball-pin is mainly subjected to sideways loading. The connection between the ball-pin and its steering lever is effected through a tapered shank and hole, acting in conjunction with a threaded portion and clamping nut.

In its once traditional form for car applications, the spherical surface of the ball-pin was embraced by an opposed pair of 'fixed' and 'floating' seatings located within the ball-joint socket. These seatings were usually spring-loaded together by means of a coil spring compressed between the floating seating and the socket end-cap. The latter often took the form of a screwed adjusting plug, so that the

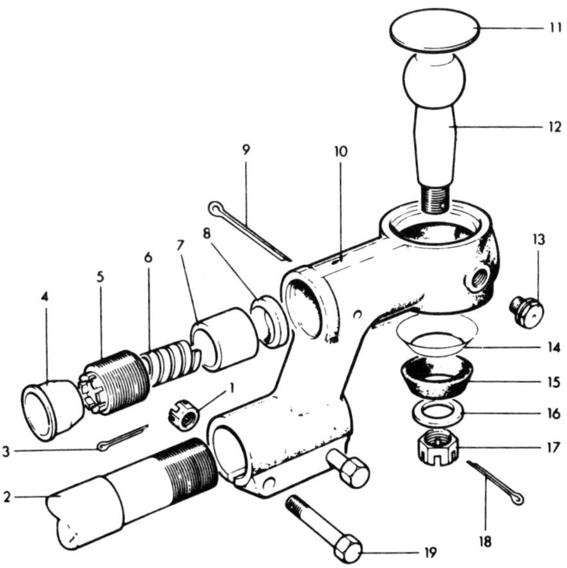

1 Nut
2 Split pin
3 Sealing cap
4 Dirt excluder
5 Adjusting screw
6 Spring
7 Ball cup — inner
8 Ball cup — outer
9 Split pin
10 Body
11 Plug
12 Ball-pin
13 Grease nipple
14 Metal cover
15 Rubber cover
16 Washer
17 Nut
18 Split pin
19 Bolt

Figure 7.10 Exploded view of pendant-type track-rod end (*Seddon Atkinson*)

build sequence was first to tighten fully the plug to 'solidify' the joint and then unscrew a fraction of a turn to provide the joint with both freedom of movement and a certain degree of frictional resistance.

A metal-to-metal combination was originally employed for the ball-pin and its seatings with provision being made for their periodic relubrication in service. From the late nineteen-fifties moulded plastics materials became increasingly wider used for the seatings and generally eliminated the need for any further lubrication after initial build.

Some ten years later a different form of ball-joint construction, incorporating new concepts in materials and design, became established to avoid the complication of spring-loaded seats. With joints of this type the ball-pin is made of constant spherical radius and is encapsulated in a partially enveloping plastics bushing. This is typically produced from high-density polyethylene material, which is resistant to abrasion, provides a smooth, low-friction bearing surface, and is self-lubricating. The bushing is installed under pressure into the tapered bore of the socket and the natural compressibility of the plastics material is relied upon to maintain the required degree of frictional resistance to joint movement. Another form of transverse steering ball-joint sometimes used in heavy commercial vehicle practice is that known as the 'pendant' joint, the distinguishing feature of which is that the fixed and floating metal seatings are disposed normal to, rather than co-axial with, the axis of the ball-pin. (*Figure 7.10*).

7.3.2 Axial ball-joint

This type of ball-joint (*Figure 7.11*) is usually confined to applications where the track rods articulate directly from each end of a rack-and-pinion steering gear, as described in § 7.6 of *Vehicle Technology 1.* Its construction is such that it is integrated with the steering rack, each end of which is counterbored to receive the spring-loaded inner seating. The outer seating is formed integrally within a steel thimble housing screwed onto the threaded end of the rack. A convenient means of adjusting ball-joint friction is thus afforded by the threaded thimble, which can then be retained by a locking collar. Alternatively, the threaded thimble may be screwed fully home against a hexagonal flange (for a holding spanner) and the ball-joint friction adjusted by selective shimming between the inner seating and its abutment in the rack.

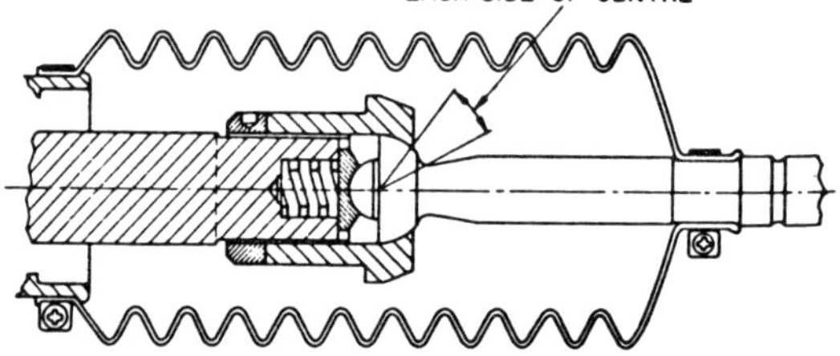

Figure 7.11 Axial ball-joint used with rack-and-pinion steering gear (*Cam Gears*)

The amount of friction torque present in the ball-joint should, of course, be strictly in accordance with the manufacturer's recommendation. In practice this is typically such that whatever position the track rod is swung to, it will not quite move under its own weight.

7.3.3 Suspension ball-joints

Since the late nineteen-fifties spherical bearings have come to be used almost exclusively for the outboard pivot connections of independent front wheel suspension linkages. This followed from a few earlier designs that were inspired by the example set by the French Citroën Company some 20 years previously, where an almost inevitable requirement for this type of suspension joint arose because of their front-wheel-drive arrangements.

As compared with the once-conventional arrangement of trunnion-mounted yokes with separate stub axles hinged from king-pins, the use of spherical bearings or 'ball-joints' that accommodate both suspension and steering movements of the yokes offers the following advantages:

(1) Simpler construction reduces unsprung mass of suspension system and thus contributes to improved road-holding.
(2) Provides improved support for the wheels, because the vertical distance separating their pivot points is always at a maximum.
(3) Front-wheel-drive universal joints can be accommodated between the span of the suspension ball-joints.
(4) Readily allows fore-and-aft compliance movements of the wheel-locating links, which reduces low-speed harshness in the ride.
(5) Enables the inboard pivot axes of the suspension links to be angled relative to each other, in side view, so as to reduce nose-diving of the car during braking.

Although the above points have been made relative to unequal transverse-link IFS systems, similar considerations apply to transverse link-and-strut installations.

7.3.4 Types and application of suspension ball-joints

Two basic constructions are used for suspension ball-joints, as follows:

(1) *One-piece* in which both suspension and steering movements of the yoke are accommodated by a combination of rocking and partial rotation of the ball-pin within its seating.
(2) *Two-piece* where the suspension and steering movements of the yoke are accommodated separately by means of a part-spherical bushing drilled to receive a rotatable thrust pin.

An important consideration in the application of ball-joints to unequal transverse link IFS systems is that the joint through which the suspension spring load is transmitted should not impose undue frictional resistance to steering movements of the yoke. For this reason either an upstanding one-piece or a downhanging two-piece ball-joint is used at the outer end of the lower link, against which the suspension spring is normally arranged to act. The former choice of transmitting the spring load via the crown of the ball was a Jaguar innovation of the late nineteen-forties which allowed a reduction in effective friction radius at the bearing seating to minimise steering effort. A similar advantage is gained by using a downhanging two-piece joint at this station and originally represented the American approach to minimising steering friction, while at the same time transmitting spring load (*Figure 7.12*).

An upstanding one-piece ball-joint is conventionally used at the outer end of the upper link. This may also be retained for the less

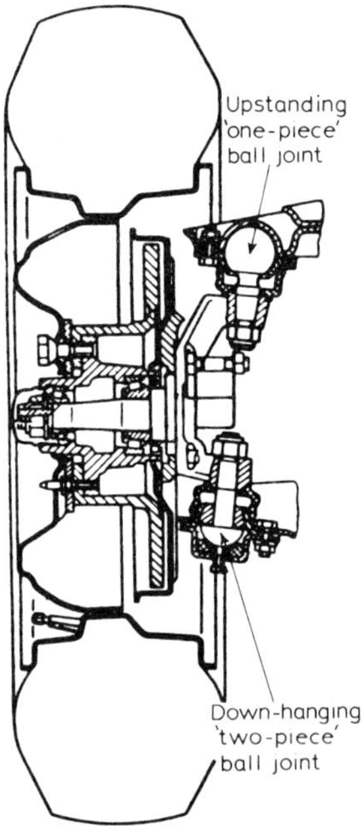

Figure 7.12 (*left*) Two types of ball-joint used with transverse-links (*Lada*)

Figure 7.13 (*below*) An example of ball-joint used with transverse link-and-strut IFS (*Nissan*)

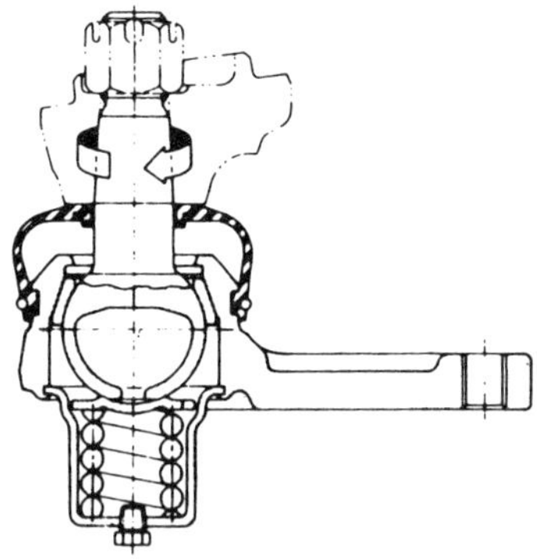

common type of layout where the suspension spring acts against the upper link, since again the spring load will be transmitted via the crown of the ball in the interests of reducing steering friction.

Early designs of suspension ball-joint were usually made adjustable either by means of a screwed cap, or selective shimming against a fixed cap, which set the endwise positioning of the 'floating' seating. Like their steering ball-joint counterparts, modern suspension ball-joints are rarely provided with means for adjustment in service.

7.4 BALL-JOINT LUBRICATION AND SEALING

From the descriptions already given of car steering and suspension ball-joints, it should be evident that in modern practice they generally require neither adjustment nor lubrication during their service life. If the latter is required it is usually at intervals of not less than 12 000 miles or 20 000 km, unless operating conditions are particularly severe, although with some older-established models this interval may be reduced to 6000 miles or 10 000 km.

A major contribution to this desirable state of affairs lies in the more dependable sealing arrangements of the modern ball-joint, which is positively sealed by a flexible 'boot' made from rubber or rubber-like plastics materials. An encircling garter spring is typically used to retain the sealing boot in a groove around the socket housing. The neck of the boot must, of course, allow for freedom of rotational movement of the ball-pin relative to the socket housing, so it therefore floats against a sealing face machined on the ball-pin supporting member. It is also good practice periodically to examine all ball-joint sealing boots and gaiters for splits or cuts and, of course, to renew them if necessary.

The steering ball-joints of commercial vehicles are either regularly greased under a planned lubrication programme, or receive oil from an automatic chassis lubrication system. Grease-lubricated joints usually feature a large capacity grease reservoir in their end-cap, a grease distributing slot across the ball that also prevents internal pressurisation of the joint, and a sealing action that permits excess grease to escape thereby preventing rupture of the boot.

As an example of routine maintenance requirements, the heavy commercial vehicles manufactured by Seddon-Atkinson require their steering ball-joints to be lubricated after the first 1000 miles or 1600 km of vehicle life and thereafter every 1500 miles or 2400 km or weekly. The grease used is to be a multi-purpose lithium-based compound whose characteristics combine water resistance with extreme pressure-and corrosion-inhibiting properties. If the vehicle is operated under conditions in which corrosive elements such as salt etc. are encountered, then a suitable calcium—lead-based grease should be used.

7.5 WEAR IN THE STEERING MECHANISM

Wear in the steering mechanism proper, that is the steering gear and its ball-jointed linkage to the road wheels, can be either wholly or partly responsible for a variety of steering complaints that can make driving both difficult and hazardous. For example, the presence of

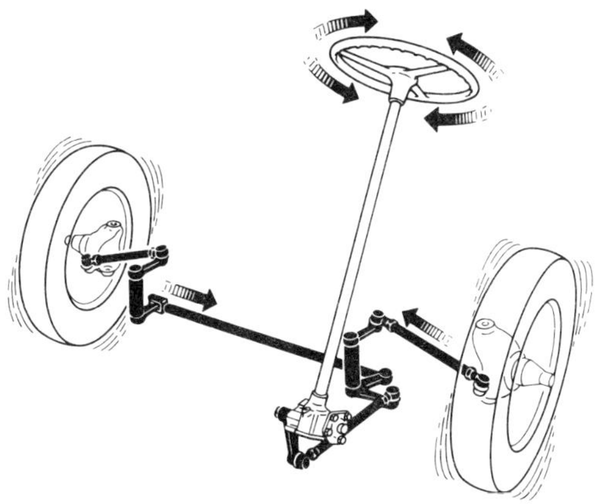

Figure 7.14 The nature of steering 'shimmy' (*Alfa Romeo*)

wear in the steering linkage can materially contribute to low-speed steering wobble or 'shimmy' of the front wheels. This may be described as a rapid oscillation of the front wheels, which try to turn in and out alternatively, causing the steering linkage to shake side-to-side (*Figure 7.14*).

7.5.1 Inspecting for wear in the steering mechanism

Clearly a process of elimination must be adopted to establish whether any free-play is confined either to the steering linkage or to the steering gear, or is cumulative to both.

Steering linkage: This may be checked for wear by jacking up the front end of the car onto axle stands placed under the suspension links so that they do not droop to the rebound position. In other words, the angularity of the steering track rods remains the same as for the normal ride height of the car. Then by grasping both front tyres at hub level and attempting to force the wheels apart and together again, the presence of any free-play in the linkage will be shown up by relative lateral movement between the fronts of the wheels.

Steering gear: A visual check on whether the steering gear requires adjustment, or possibly replacement, can be made with the car on a ramp. The drop-arm may then be observed for follow-up movement, while an assistant 'waggles' the steering wheel either side of the straight-ahead position against the turning resistance of the tyres.

So far as the Department of Transport's inspection of goods vehicles is concerned, if a point on the rim of the steering wheel moves without the road wheels turning, for a distance more than one-fifth of the diameter of the steering wheel, then this can be a reason for rejection in the case of non-power-assisted steering mechanisms. For example, there should not be more than about 105 mm free-play on a 525 mm diameter steering wheel. In the case of cars and light goods vehicles not exceeding 1525 kg, free-play at the steering wheel in excess of 75 mm is considered unacceptable where worm-and-follower steering gears are employed, and only one-sixth of this amount is permitted for the more direct-acting rack-and-pinion steering gears.

Once having pin-pointed the source or sources of free-play in the steering mechanism it is then, of course, necessary to carry out a more detailed examination of the affected parts. From this it may be decided by a combination of experience and knowledge of the particular manufacturer's service techniques whether replacement, simple adjustment or overhaul is required.

8 Car and commercial vehicle brakes

8.1 THE ACTION OF SHOE BRAKES

In § 8.2 of *Vehicle Technology 1*, the drum brake in its basic form was described and its leading and trailing shoes were identified. From this it may be recalled that a leading shoe will generate a greater braking drag force on the drum than a trailing one for the same applied load at their toes (or 'tips'). We should perhaps now express this difference in braking drag force in terms that are more familiar to an automotive engineer, by saying that a leading shoe possesses a higher 'shoe factor' than a trailing one. That is:

Shoe factor = drum drag/shoe tip load

Other things being equal, the shoe factor is about three times as great for a leading shoe as for a trailing one. Considering next the combined effect of both shoes, we can further say (except in the more complicated case of a fixed-cam shoe brake) that the 'brake factor' equals the sum of the separate shoe factors, or simply:

Brake factor = total drum drag/total shoe tip load

8.1.1 Two-leading shoe (2LS) brakes

It was almost inevitable that a demand for a brake factor higher than could be obtained from a simple leading and trailing shoe (L/TS) brake should arise first in the field of motor-racing. As a result we find that a further development of the drum brake, known as the 'two-leading shoe', or 2LS, type, was first adopted by the German firms of Daimler-Benz and Auto-Union for their legendary racing cars of the nineteen-thirties.

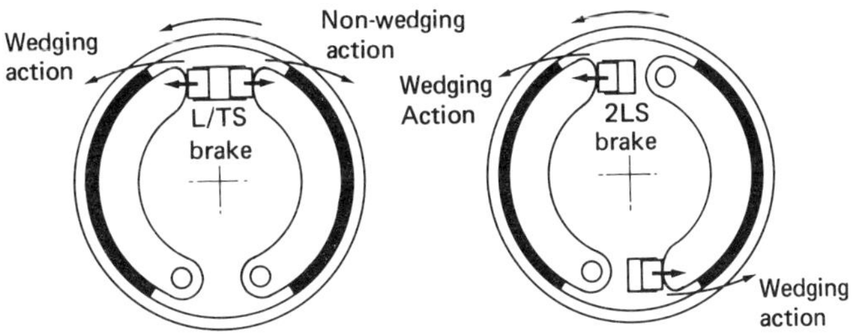

Figure 8.1 The basic difference between leading-and-trailing (L/TS) and two-leading shoe (2LS) brakes

Following the introduction of 2LS front brakes on the 1940 production model Chrysler cars in America, this particular arrangement of front brakes became increasingly popular in the United Kingdom and on the Continent, although it was not further pursued in America. It had the merits of concentrating the greater braking torque on the more heavily loaded front wheels, while allowing the retention of L/TS rear brakes which were equally effective in the forwards and reverse directions and therefore better suited for handbrake operation. For cars the 2LS brake was largely superseded by the disc brake in the early nineteen-sixties, but continues to be used on medium-sized commercial vehicles not only for the front but also for the heavily-laden rear wheels.

8.1.2 Types of 2LS brake

The identifying features of any 2LS brake are that the shoes pivot from diametrically-opposite abutments; each shoe is provided with a separate means of actuation; and that during forwards braking the direction of drum rotation is from the toe to the heel of each shoe.

We next have to consider the different means by which the two shoes can be separately actuated, these being as follows:

(1) Hydraulic.
(2) Mechanical.

Hydraulic 2LS (*Figure 8.2*): This is the original and simplest form of 2LS brake. Each shoe is actuated by a separate, single-acting, hydraulic wheel cylinder of which the blind end forms an abutment for the adjacent heel of the opposite shoe. The two wheel cylinders are linked on either side of the brake backplate by a separate pipe and the brake hose connects to the upper cylinder.

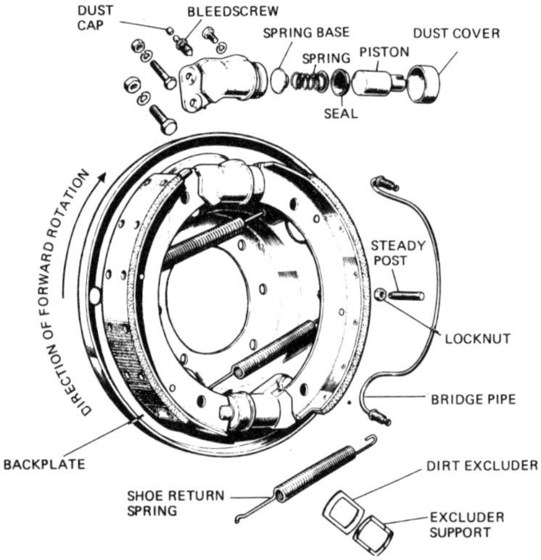

Figure 8.2 Hydraulically-operated 2LS commercial-vehicle brake (*Girling*)

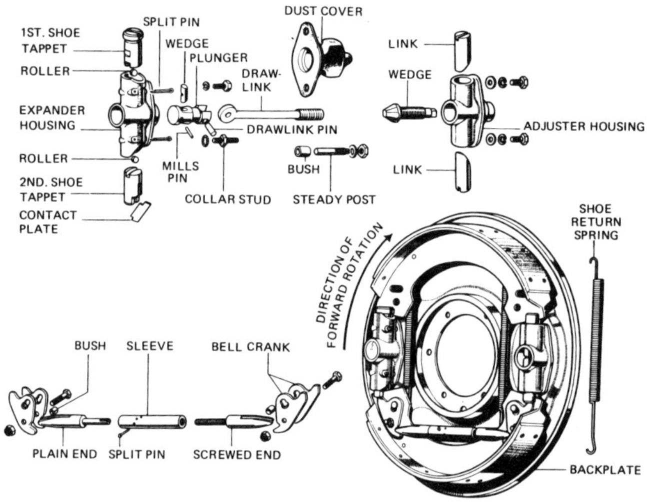

Figure 8.3 Mechanically-operated 2LS commercial-vehicle brake (*Girling*)

Mechanical 2LS (*Figure 8.3*): For this arrangement a 'wedge and tappets' expander is used in conjunction with a 'bellcranks and strut' linkage on what would normally be the trailing shoe of a shoe brake. The trailing shoe is therefore converted into a second leading shoe (*Figure 8.4*), because the appropriate tappet of the wedge expander does not bear directly against the shoe tip, but acts through the bell cranks and strut linkage to expand the other end of the shoe.

A perhaps simpler mechanical arrangement for actuating two-leading shoes is that found in the front brakes of some motor-cycles, where the shoes are expanded by two diametrically-opposite cams. The camshafts are linked externally by a levers and pull-rod mechanism, one lever being extended for connection to the cable actuating system.

8.1.3 Advantages and disadvantages of 2LS brakes

As compared to a L/TS brake the 2LS brake offers the following advantages:

(1) Greater braking torque for the same operating force or, perhaps more realistically, a smaller operating force to achieve the same braking torque.

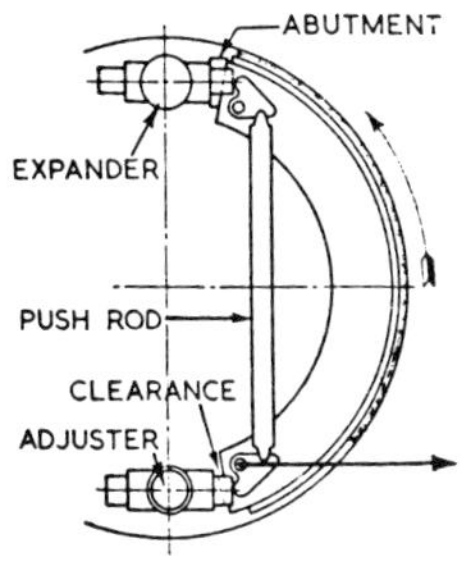

Figure 8.4 Trailing shoe converted to leading shoe by bell-cranks and strut

(2) Since each shoe is equally loaded there are no unbalanced loads imposed on the wheel bearings.

So far as disadvantages are concerned these are:

(1) When braking in the reverse direction the two leading shoes become two trailing shoes, so that a considerably greater operating force is required to generate the same braking torque.
(2) The 2LS brake is very sensitive to changes in friction level between the linings and drum, so that as the brake heats up there is a greater falling off in brake torque ('fade') for a given operating force.
(3) The purely hydraulic version does not lend itself to additional operation by the handbrake system, but this can readily be arranged in the mechanical version through the wedge expander mechanism.

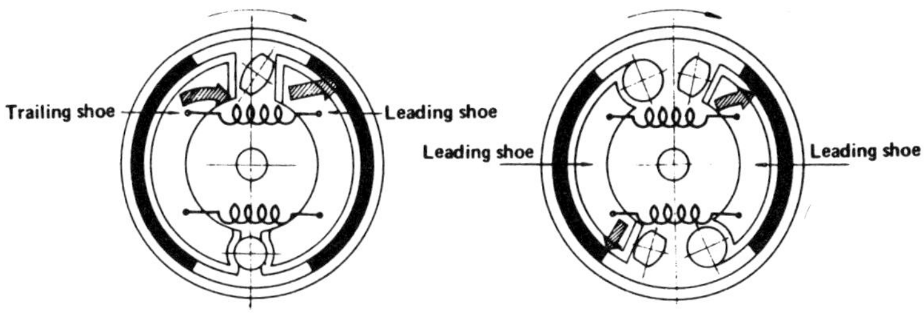

Figure 8.5 Comparison between cam-operated L/TS and 2LS motor-cycle brakes (*Yamaha*)

Regarding the second disadvantage, this deserves a few words of simple explanation. As mentioned in § 8.2.1, brake fade arises from a reduction in the frictional properties of the lining material as its temperature increases. This in turn has the effect of altering the direction of the resultant force on the brake shoe and hence the distance it acts from the shoe pivot. Since this leverage acts in

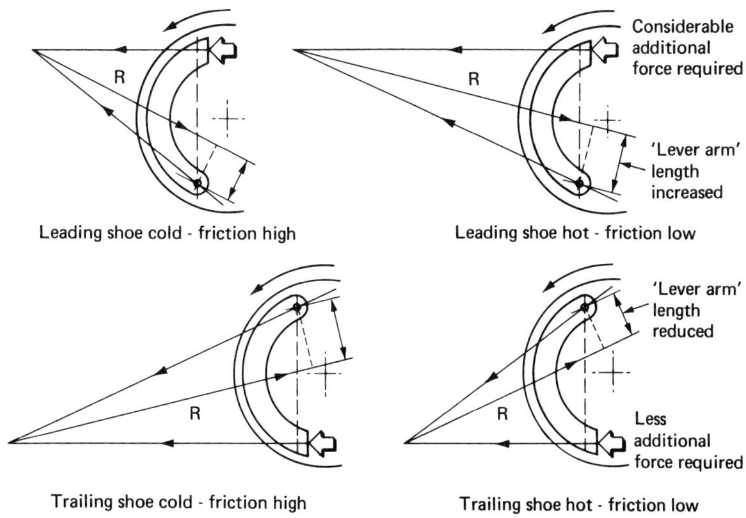

Figure 8.6 Simplified representation of additional operating force required for L/T shoes with falling friction

opposition to that of the brake shoe apply force, which also acts about the shoe pivot, then clearly with the onset of brake fade the brake apply force must be increased to maintain the same braking torque. The reason why fade does not greatly increase the effort required to operate a trailing shoe, in comparison with a leading shoe, is that in the former case fade decreases the effective lever arm of the resultant force about the shoe pivot, whereas with a leading shoe it increases the length of the effective lever arm. In other words, the operating force on the leading shoe must be increased to a greater extent to maintain a given brake torque. It is, of course, only fair to add that the force needed to operate a trailing shoe is greater in the first place.

8.2 BRAKE-LINING MATERIALS

Reducing the speed of a vehicle requires the addition of a retarding force, applied conventionally through a friction braking system. This converts the kinetic energy of motion into frictional heat energy and dissipates it to the surrounding atmosphere, chiefly by forced conduction, until the vehicle comes to rest, and then by radiation in the absence of air flow over the brakes. Indeed, the brakes may be considered as being a form of heat engine, albeit in the negative sense because they convert mechanical energy into heat rather than vice versa. Life can be much more difficult for the brakes, however, since they are often called upon to decelerate the vehicle in a fraction of the time that the engine is normally expected to take to accelerate it. In other words, the amount of energy handled by the brakes can be the equivalent of many times the power developed by the engine.

8.2.1 Basic requirements of a modern lining material

For an acceptable friction brake material these may be summarised, not necessarily in order of priority, as follows:

Resistance to wear: To the vehicle user and also the tribologist who studies friction and wear, this requirement is perhaps the most important. A friction material should be hard enough to give a satisfactory wear life, but soft enough to avoid damage to the mating surface of the drum or disc. The most significant factor governing wear is temperature, since the rate of wear increases disproportionately with it.

Correct friction level: The friction coefficient of most modern brake lining and pad materials against cast iron normally lies between 0.3 and 0.5. It should be sufficiently high to avoid excessive operating forces in the braking system, but not so high as to cause 'grabbing' of the brakes. In any event the friction level must be high enough for the brakes to provide a minimum retardation sufficient to meet the Department of Transport requirements.

Resistance to heat fade: This is a measure of how well the friction level of the material is maintained despite an increase in the braking temperature. Although the asbestos content of the material is little affected by heat, it is the resin used for bonding the material together that can melt and smear itself between the rubbing surfaces and thereby reduce the friction level. For a good quality lining or pad material this will return to its original value on cooling, even after being subjected repeatedly to severe heating, but for a less acceptable material the recovery from fade will be poor.

Minimal friction variations: The lining or pad material should be relatively insensitive to variations in operating pressure and rubbing speed. In these respects a less acceptable material would exhibit a reduced friction level at high operating pressures and at high rubbing speeds, neither characteristic being conducive to safe high-speed driving.

Reasonable wet behaviour: It is important that linings or pads should recover their original friction level quickly after exposure to water and should not behave erratically during the drying out process. The more porous a material is then the longer it will take to recover.

Quietness of operation: Nowadays this is an important comfort factor in vehicle operation. In combination, the friction lining material and the structural features of a brake assembly should contribute to an operational noise level that remains below the audible.

8.2.2 A typical modern brake pad material

General description of Ferodo 2424F: This is a rigid moulded, non-metallic friction material, medium brown in colour, having a random-fibre asbestos base. It is one of the high-friction members of the Ferodo Two-Four series of disc pad lining materials, having also

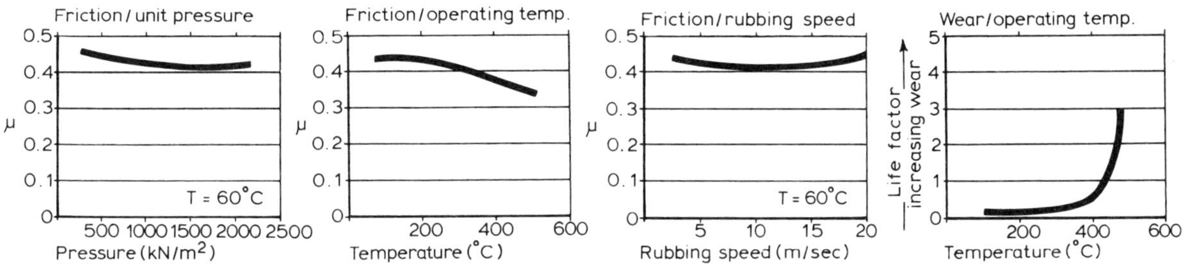

Figure 8.7 Operating characteristics of Ferodo 2424F disc pad material

good durability and satisfactory fade resistance together with freedom from noise. For design purposes its coefficient of friction μ is quoted as 0.42. The operating characteristics of this material are shown in *Figure 8.7.*

As in the case of clutch friction linings, the present trend is toward the use of asbestos-free brake materials, which typically include synthetic material, steel or glass fibres in their composition!

8.3 WEARING OF BRAKE LININGS AND DRUMS

Before discussing the factors that influence the wear life of brake linings and drums, it is perhaps only logical to enquire the actual nature of 'wear'. In fact this is a topic that has long engaged the activities of research workers, since wear may take several forms dependent on operational and environmental circumstances. However, the form of wear with which we are concerned is often referred to as a 'ploughing' action and will now be briefly explained.

Although a function of the brake lining is to distribute load evenly over the surface, it can only do this in the general sense of the description, because in reality the rubbing contact between lining and drum takes place over a whole series of tiny high spots (this would still be true even for precision-ground, flat metal surfaces). The increase in pressure at these high spots is such that they momentarily weld themselves together, but as long as there is relative motion between the drum and linings these tiny welded junctures are continually being broken—thus the 'ploughing' action. It is, of course, this action that produces the friction force and also causes some damage to the rubbing surfaces, or in other words what we recognise as wear.

8.3.1 Factors influencing the wear life of linings

These may be summarised as follows:

Temperature attained: High rubbing speeds and contact pressures, either singly or in combination ('heavy braking'), will increase operating temperatures and have the greatest effect on reducing the wear life of the linings
Mating surface condition: To reduce wear the rubbing surface of the drum should have as fine a finish as possible and its material be resistant to scoring from the wear products of the lining material.
Operating environment: The ingress of muddy water from wet roads can lead to corrosion of the rubbing surface of the drum and will increase the rate of wear on the linings.

8.3.2 Recognising brake lining and drum wear

Clearly this is a topic that better lends itself to illustration rather than description in words. *Figure 8.8* shows two examples of brake neglect in service from the Ferodo 'chamber of horrors' collection.

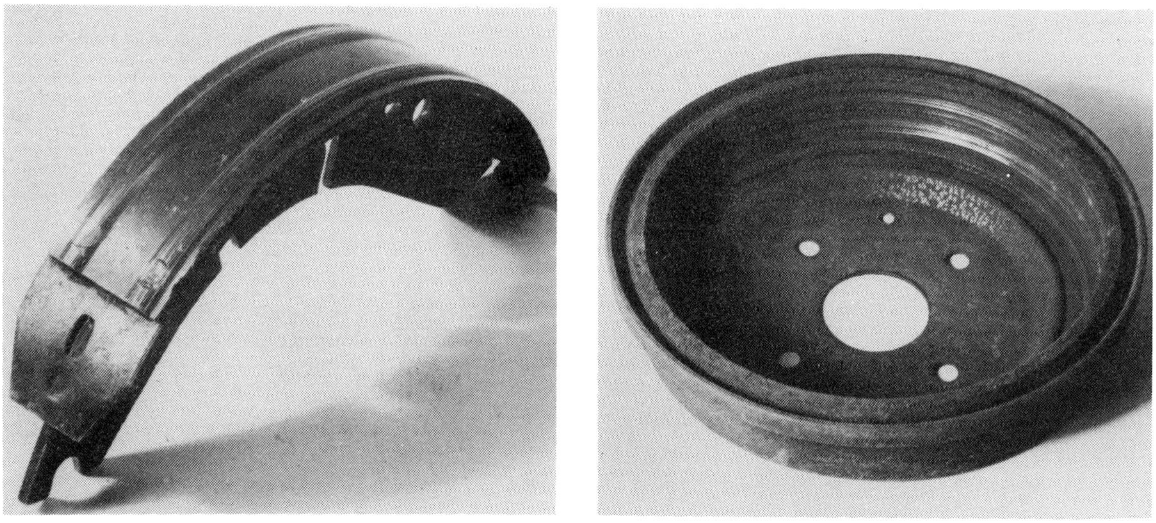

Figure 8.8 Drum brake neglect in service (*Ferodo*)

8.4 DIAGNOSING DRUM-BRAKE WEAR

The various conditions of lining and drum wear shown in *Figure 8.9* may be diagnosed as follows:

Scoring: Three possible causes of this are abrasive debris embedded in the lining material; new linings mated with already scored drums; and the drum material too low in hardness for the lining material.
Crazing: This is caused by repeated overheating and cooling of the brakes due to heavy usage and appears as randomly-orientated cracks on both the friction track of the drum and the lining material.
Heat spotting: Either the lining is not sufficiently conformable to the drum, or the latter is distorted so that lining contact occurs only at small heavily-loaded areas. Discoloured regions and sometimes cracks appear on the friction track of the drum, this being accompanies by heavy gouging of the lining.

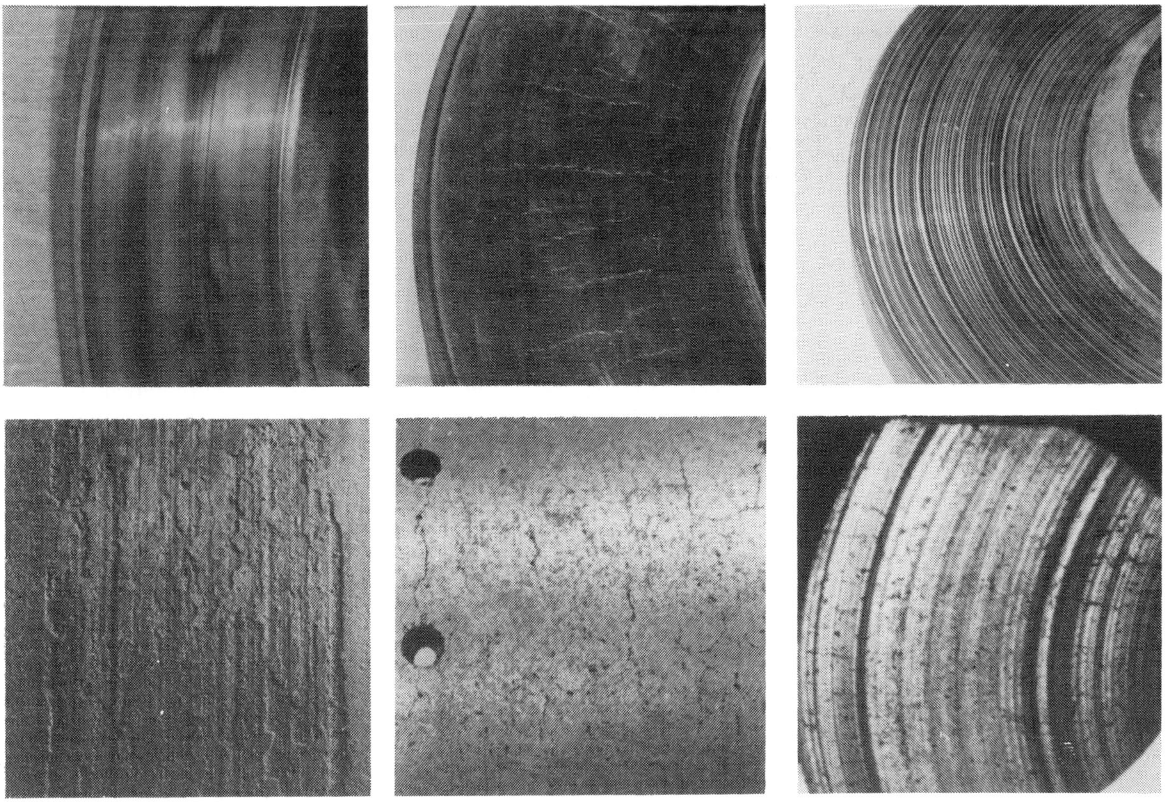

Figure 8.9 Recognising drum and brake lining wear. Upper line — metal surfaces, lower line — friction material surfaces. (*left*) Heat spotting, (*centre*) crazing, (*right*) scoring (*Ferodo*)

Fading: Where this has occurred the lining material degrades or flows at the rubbing surface, an explanation of heat fade having been given earlier in § 8.2.1. Surface cracks on the brake drum friction track again may be associated with such arduous usage, their presence or otherwise being influenced by the metallurgical specification of the cast iron. Apart from heavy usage of the brakes, fade can also be caused by the brakes dragging on.

8.5 HANDBRAKE ARRANGEMENT FOR DISC BRAKES

The addition of a handbrake mechanism to the disc brake poses certain problems, which partly account for the mixed system of front disc and rear drum brakes being used in conventional practice, although an alternative approach is to build a small drum handbrake mechanism into the disc hub.

Problems arise because the disc brake is to all intents and purposes non-self-energising in its action and does, in fact, have the lowest brake factor of all automotive brakes. Hence, it follows that the force required to operate it is necessarily the highest. So far as the handbrake system is concerned, this implies that a large leverage must be provided between the driver control and the disc pads. Therefore a small amount of wear at the pads can be magnified into a greatly increased travel at the handbrake lever. It is for this reason than an automatic adjusting device is generally specified for such systems. Also, the operating linkage must be made as rigid as

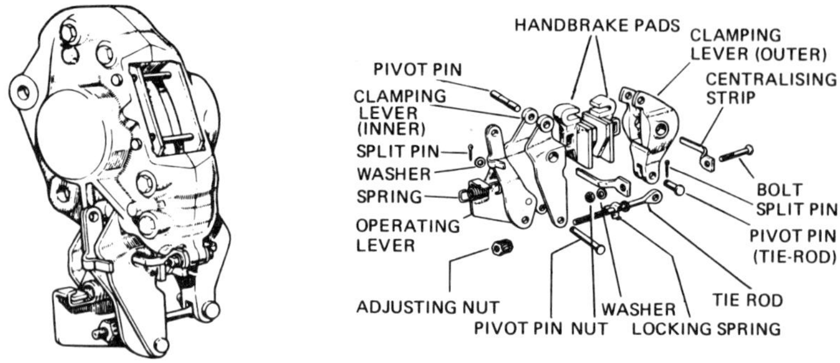

Figure 8.10 Manually-adjusted disc handbrake (*Girling*)

possible, because any unwanted flexing would again increase the travel at the handbrake lever. As a knowledge of automatic brake adjusting devices is not required until a later stage, the following description is of a manually-adjusted disc handbrake (*Figure 8.10*).

The brake caliper is provided with an auxiliary pair of small wedge-shaped pads, which are brought into contact with the disc by a downhanging linkage connected to the handbrake system. An upstanding operating lever is pivoted from the lower end of the longer of two pad clamping levers and is connected to the shorter one by an adjustable tie-rod. When the handbrake is applied the operating lever is pulled away from the caliper body, so that one clamping lever and pad is moved towards the disc by the tie-rod and the other by the reaction at the pivot for the operating lever. The very small clearance between each pad and the disc is set by the adjusting nut on the tie-rod. Centralising strips are sometimes incorporated to retract the clamping levers and pads when the handbrake is released.

8.6 THE ACTION OF DISC BRAKES

It is now a matter of motoring history that the 'spot-type' of disc brake (so-named to distinguish it from a much earlier and less successful type of disc brake that acted in the manner of a plate clutch) contributed greatly to the motor-racing successes of Jaguar cars in the mid-nineteen-fifties. Following this promising initial development by the Dunlop Company the disc brake became increasingly adopted, at least for the front wheels, on most production cars.

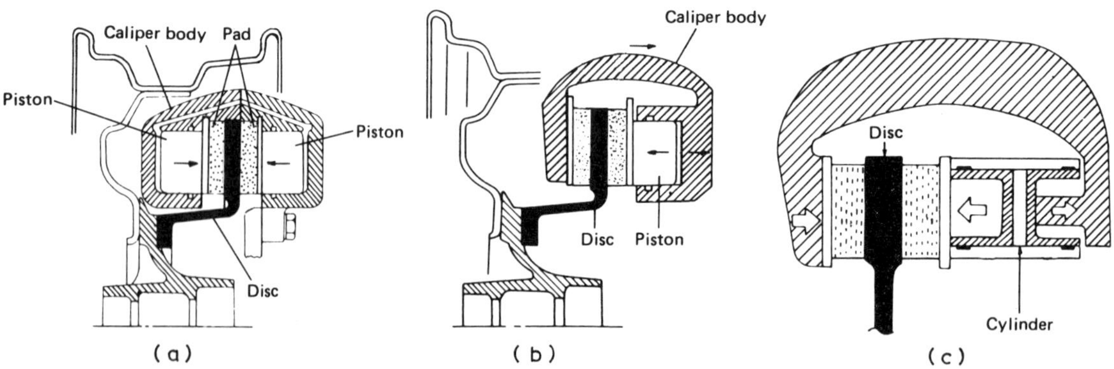

Figure 8.11 Types of non-pivoting disc brake. (a) Fixed caliper and cylinders, (b) floating caliper and cylinder, (c) floating caliper and fixed cylinder (*Suzuki*)

The advantages of the disc brake over the drum brake are generally as follows:

(1) Less prone to heat fade because a higher rate of cooling is achieved with the rubbing surfaces of the disc being exposed to the air stream.
(2) Any heat fade that does occur is less pronounced because being non-self-energising reduces its sensitivity to changes in friction level.
(3) Inefficiency and apparent fade caused by expansion of the drum, and consequent poor bedding of its shoes, are absent in the case of the disc brake.

The basic operating principle of the disc brake was described in § 8.2.3 of *Vehicle Technology 1,* so it is now proposed to classify the different forms a disc brake may take and to indicate their relative merits, as below:

(1) Fixed caliper and cylinders.
(2) Floating caliper and cylinder.
(3) Floating caliper and fixed cylinder.

8.6.1 Fixed caliper and cylinders

This represents the earliest development of the modern disc brake and remained the most widely-used type until the early nineteen-seventies (*Figure 8.12*). It comprises simply a mounting half and a rim half, which are bolted together and straddle the brake disc rigidly attached to the wheel hub. One or more cylinders in each half of the caliper house sealed and self-adjusting pistons. These press the friction pads against the respective sides of the disc, each pad being located in a recess within the caliper. When the brake is applied both pads are moved simultaneously into contact with the disc, because a hydraulic pressure balance between the opposed cylinders is obtained either by an external bridging pipe or internal drillings through the caliper bridge itself.

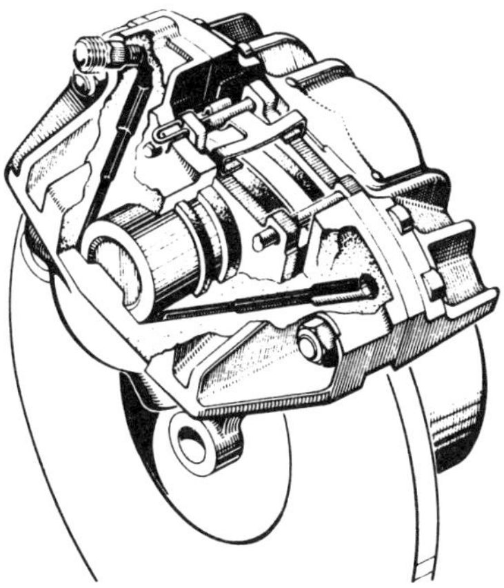

Figure 8.12 Cut-away view of fixed caliper and cylinders disc brake (*ATE*)

The drawbacks generally associated with this form of disc brake may be summarised as follows:

(1) Since the rim half of the caliper houses one or more pad operating pistons, its bulk can be difficult to accommodate within the confines of the road wheel.
(2) Under particularly severe usage the temperature reached by the brake fluid, especially in the transfer passage between the caliper halves and in the outboard cylinder(s) shrouded by the

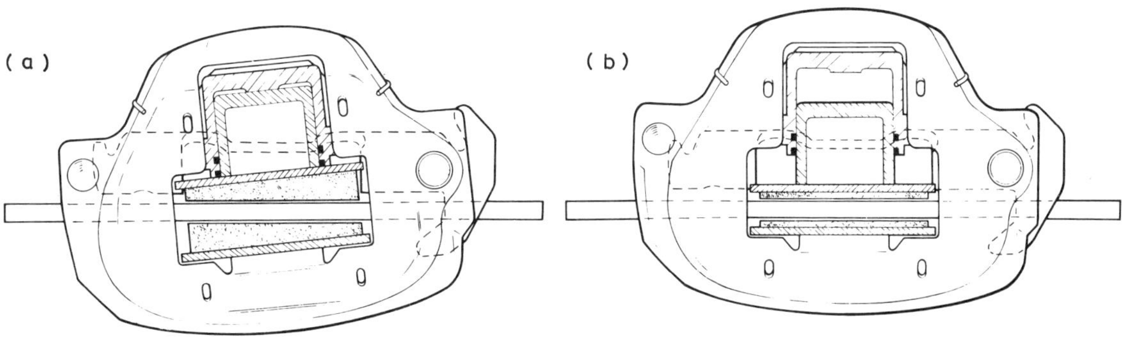

Figure 8.13 Pivoting caliper and cylinder disc brake. (a) Assembly with new pads, (b) pads fully worn (*Lockheed*)

wheel, may exceed its boiling point so that vaporisation and possible loss of braking could occur.

(3) It is difficult to provide an integral handbrake mechanism with this form of disc brake and, as previously described, an auxiliary pair of pads with their own actuating linkage is required.

To their credit we can say that:

(1) Virtually all the movable parts are lubricated by hydraulic brake fluid.
(2) They offer the simplest form of pad location.
(3) Four-cylinder versions readily lend themselves to dual hydraulic circuits for safer braking.

8.6.2 Floating caliper and cylinder

The distinguishing feature of this type of disc brake is that only the inboard pad is directly operated by hydraulic means, the reaction force from this being transmitted to the outboard pad by an abutment in the floating caliper. A floating action for the caliper is obtained by virtue of either a pivoting or a sliding mounting system.

In one established example of the pivoting mounting system, the caliper takes the form of a frame produced from a rigid steel pressing. A single hydraulic cylinder and piston is secured on the inboard side of the caliper frame, which is pivoted at its upper end to the mounting bracket. Therefore when the inboard pad is pressed into contact with the disc by the hydraulic piston, the caliper frame and cylinder swings in opposition and applies the other pad with equal pressure. As the pads wear the angular relationship between the caliper frame and disc gradually changes, so to compensate for this the friction material is initially of tapered section, but approaches parallelism with the disc as wear proceeds.

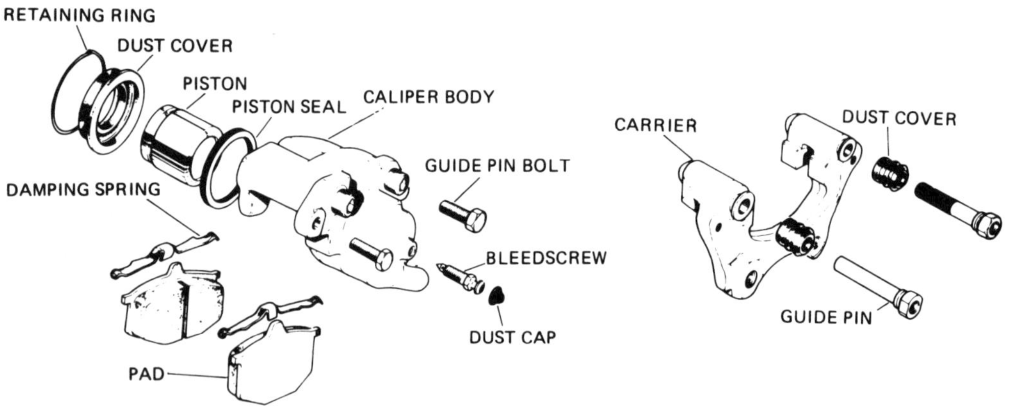

Figure 8.14 Exploded view of a modern floating caliper and cylinder disc brake, showing carrier bracket for reacting brake torque (*Girling*)

Current developments of the disc brake would seem to be concentrated on the sliding mounting system for the caliper. In a typical modern example a 'fist'-type sliding caliper is used, this being so-called because of its similarity in form to a partly clenched fist. A hydraulic cylinder and piston (in some cases two are used) is housed in the wrist portion and presses the inboard pad into contact with the disc, while the fingers form the moving abutment or claw for the outboard pad. A carrier bracket reacts brake torque and provides the necessary support and location for the pads. Also, it is furnished with precision-bored guides to receive the mounting pins of the sliding caliper, one of the pins being rubber sleeved to prevent rattle in certain types of Girling Colette brake.

It should be appreciated that the sole purpose of the sliding caliper is to clamp the pads onto the disc, since it plays no part whatsoever in reacting the drag forces on the pads when the brake is applied, this being the function of the carrier bracket mounted from the steering yoke. The sliding surfaces of the caliper mounting pins are greased and sealed to minimise friction; otherwise problems can arise such as unequal pad apply forces, false self-adjustment and dragging of the brake.

The advantages of the modern floating caliper disc brake may be listed as follows:

(1) Brake fluid operating temperatures can be reduced by virtue of the more remote and unshrouded inboard mounting for the hydraulic cylinder(s).
(2) The absence of hydraulic cylinders on the outboard side of the disc makes for easier accommodation within the confines of the road wheel.
(3) A sliding version of floating caliper readily lends itself to the addition of an integral handbrake mechanism.

On the debit side we may include the following points:

(1) Pad replacement can be less easy with a sliding version of floating caliper, because the fist must be either completely or partially removed from its mounting bracket.
(2) There is a risk of corrosion and possible seizure in designs where the caliper floats on open slides, which gives rise to the operational problems mentioned earlier.

8.6.3 Floating caliper and fixed cylinder

This design of disc brake represents something of a compromise between the 'fixed' and 'floating' caliper constructions, because although it is basically a sliding caliper type it retains the easy pad replacement of the fixed caliper. Its constuction is such that a hydraulic cylinder with opposed pistons is rigidly attached to a mounting bracket on the inboard side of the brake disc. A steel yoke or 'frame' straddles the disc and can slide in grooves provided alongside the cylinder body.

When the brake is applied, the hydraulic pressure in the fixed cylinder forces the pistons apart, one of which acts directly on the inboard pad and the other indirectly on the outboard pad via the sliding yoke. It should be noted that the latter is also responsible for transmitting the drag forces on the outboard pad to the cylinder body and mounting.

This particular form of floating caliper brake has the following good points:

(1) In common with the floating caliper and cylinder type, brake fluid temperature can be reduced.
(2) Its frame construction allows an increased diameter of disc within a road wheel of limited size.

On the other hand it could be at a disadvantage in respect of the following:

(1) The overall length of the caliper frame does not make for a compact installation of the outboard friction pad.
(2) It does not lend itself to dual hydraulic circuits for safer braking.

8.7 THE EFFECTS OF EXCESSIVELY-WORN FRICTION PADS

Although disc brakes generally lend themselves to simple servicing procedures they can nevertheless be the victims of potentially dangerous neglect. This is because, unlike many drum brakes, they are self-adjusting and can therefore function for long periods without calling attention to themselves. If the disc brake pads are allowed to wear beyond the generally recommended limit of 3 mm lining thickness remaining, the effect may be two-fold:

(1) The reduced thickness of heat insulation offered by the pad material between the rubbing surfaces and the brake fluid in the caliper cylinder(s), introduces a risk of fluid vaporisation and possible loss of braking under conditions of severe usage.
(2) In the event of the pad material wearing to the extent that its steel backing plate begins to contact the disc, then apart from scoring the latter the action of the brake will become both inefficient and noisy.

8.8 RECOGNISING EXCESSIVE WEAR AND REPLACING DISC-BRAKE PADS

Recognition of when the brake pads have reached the end of their useful life may take two forms, as follows:

(1) 'On-board' indication to the driver.
(2) 'Off-board' examination by the service engineer.

On-board indication: The driver is provided with an early-warning system of the pads wearing out by either audible or visual indication, or by change in 'feel' at the brake pedal. An audible indication may be obtained simply by adding a 'squealer' spring to the inboard brake pad. When the latter has worn to a predetermined thickness the spring becomes flush with the pad rubbing surface, so that application of the brake will cause the spring to rub on the disc and emit a high-pitched squeal. A visual indication can be arranged electrically by having a sensor embedded in the brake pad. Wearing of the pad eventually exposes the sensor, therefore when the brakes

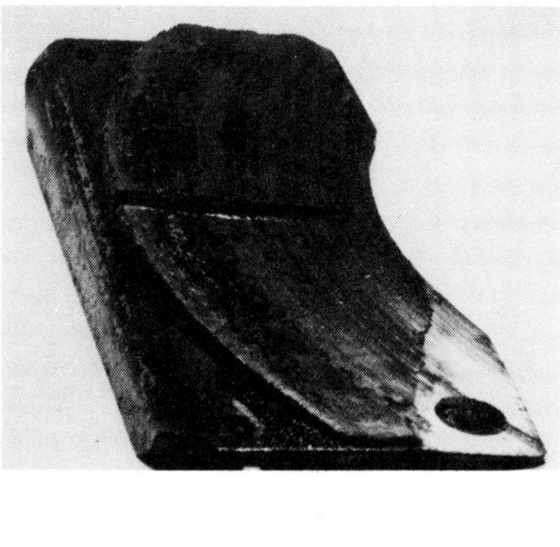

Figure 8.15 Disc pad neglect in service (*Ferodo*)

are applied an electrical circuit is completed through contact of the sensor with the disc and illuminates an indicator visible to the driver.

A change in feel at the brake pedal with worn pads can be enforced by their backing plates eventually embracing a cross-leaf spring, which also acts as a retaining device for the brake pads. Since the driver will exert a greater effort on the brake pedal in trying to overcome the blocking effect of the cross-leaf spring, the 'hard' pedal feel indicates that the pads require replacement.

Off-board indication: This involves the service engineer in following the general recommendation of brake equipment manufacturers to check visually the state of wear of the brake pads themselves at least every 5000 miles (8000 km). Each wheel must be removed in turn so that every pad can be viewed for edge thickness, since it is the thinnest one that limits the life of the set. The brake pads are also checked for excessive wear in the MOT vehicle test.

8.8.1 Replacing the disc-brake pads

As explained in § 8.6, there are several different forms of disc-brake construction, so clearly it is desirable to follow the servicing instructions of the particular manufacturer concerned until complete

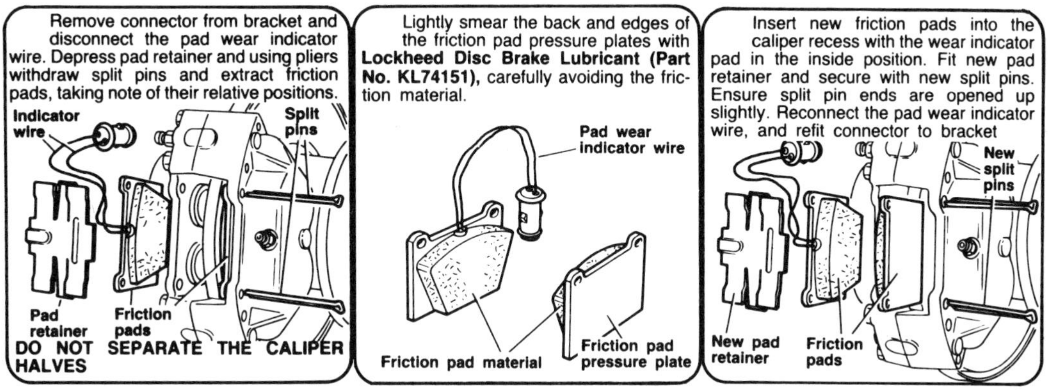

Figure 8.16 Replacing disc-brake pads (*Lockheed*)

familiarity with these units is acquired. However, the means by which the disc pads may be changed on the Lockheed Type 4 × 36 MB brake calipers fitted to the front wheels of the Austin Mini Metro 1000 and 1300 are illustrated in *Figure 8.16* as a practical example.

8.9 MANUAL ADJUSTERS FOR BRAKE SHOES

Since the adjustment facilities required by brake shoes with pivoting and sliding abutments can differ, it is convenient to consider them under separate headings, as follows:

Pivoting shoes: These are usually provided with a 'centralising' adjustment at their heels and a 'wear' adjustment at or near their toes, although both adjustments can be effected at the heels of the shoes where a wedge-type adjuster is used between them. The purpose of centralising adjusters is to ensure that during initial build and later during reline of the brake, there is no localised bedding of the liner at either the heel or the toe of the shoes. This is because a heel bedding shoe will produce an inefficient brake and a toe bedding one a grabbing brake. As its name suggests, a wear adjuster is provided for each shoe so that it can be set closer to the drum to compensate for lining wear.

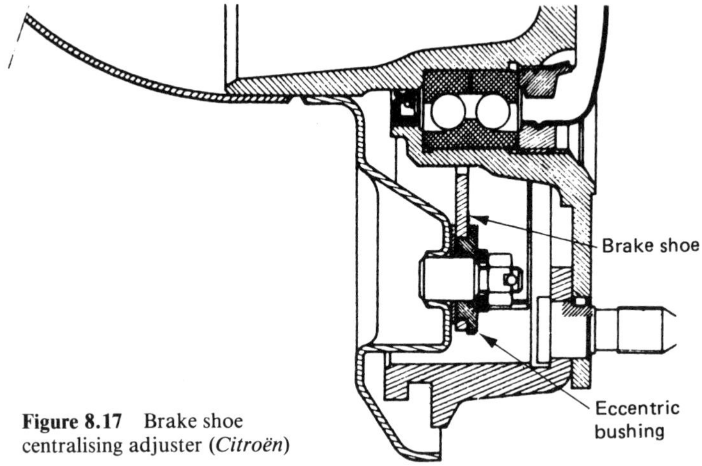

Figure 8.17 Brake shoe centralising adjuster (*Citroën*)

Sliding shoes: These were first featured in proprietary drum braking systems during the late nineteen-forties and by the nature of their mounting are self-centring within the drum, thereby eliminating any need for a centralising adjuster. A wear adjuster for each shoe is, of course, still required. The advantages of the sliding shoe over the pivoting type are generally that it allows more uniform wear along the length of the lining and it can be more easily serviced.

8.9.1 Types of adjuster

Centralising adjuster: A typical example of this type of adjuster comprises a hexagonally-headed eccentric bushing at the heel pivot of the brake shoe. Once the shoe has been centralised, the bushing is clamped to prevent further rotation, but the shoe remains free to pivot about the bushing.

Wear adjusters: Several forms of wear adjuster may be encountered,

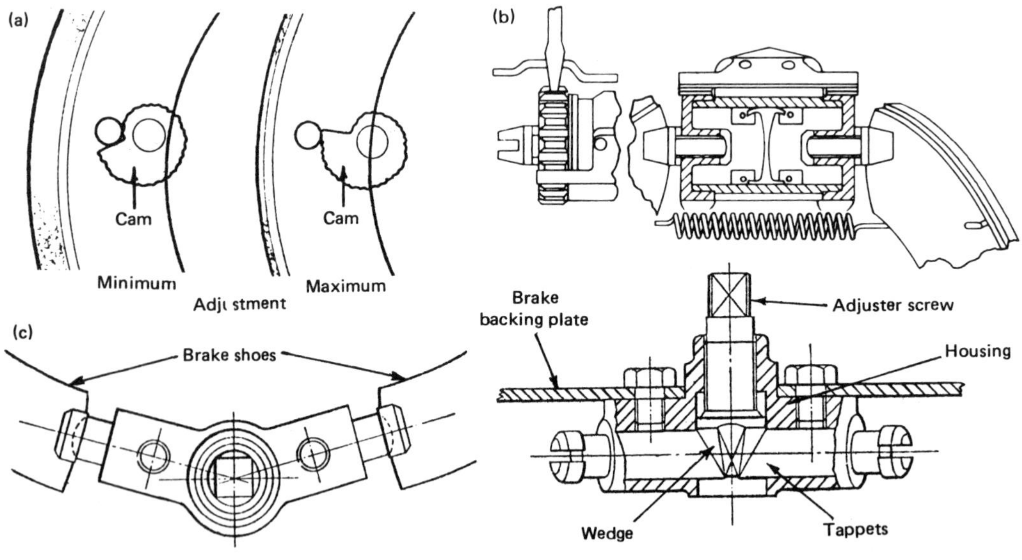

Figure 8.18 Examples of brake shoe wear adjusters. (a) Snail cam (*Girling*), (b) screwed plug (*ATE*), (c) wedge and tappets (*Girling*)

especially those known as 'snail cam', 'screwed plug' and 'wedge' (*Figure 8.18*). Snail cams are usually provided with serrated edges and may act either between the brake backplate and a striker pin on the shoe web, or between the wheel cylinder piston and shoe tip. A screwed plug adjuster may be incorporated with the wheel cylinder piston and therefore also acts against the shoe tip. In contrast, the wedge type adjuster acts between the brake backplate and the heels of the shoes and thus furnishes them with adjustable abutments. Slight re-positioning of the wedge adjuster on the backplate, before its retaining bolts are finally tightened, allows for the centralisation of pivoting-type shoes when they are expanded within the drum.

8.10 THE PROPERTIES OF BRAKE FLUIDS

Although synthetic fluids in their various formulations are no strangers to the hydraulic equipment engineer, that fluid which is used in the hydraulic braking systems of most motor vehicles has long been regarded with a certain amount of curiosity by automotive service engineers. This is probably because the manufacturers of brake fluids do not normally volunteer the chemical composition of their products, but in any event such information can be meaningful only to a lubrication chemist. However, since we are required to have some knowledge of the constituents and properties of brake fluid the following basic information should suffice:

Constituents: A conventional brake fluid includes the following chemical components:

(i) Glycols to ensure that the fluid is compatible with the rubber materials in the braking system.
(ii) Gylcol ethers to enable the fluid to flow freely and therefore have the desired viscosity.
(iii) Polyglycols to confer the required lubricity on the fluid.
(iv) Corrosion inhibitors to protect the metal parts of the braking system.

Freezing and boiling points: A typical high-quality brake fluid would remain chemically stable over a temperature range of –65°C to +220°C.
Viscosity: The maximum acceptable viscosity of a brake fluid is limited by its satisfactory behaviour at –40°C. That is, the brakes should not respond sluggishly either during application or release at this low temperature.
Specifications: Since 1945 the American Society of Automotive Engineers (SAE) has issued specifications relating to the minimum required performance of brake fluids, as also does the Department of Transportation in that country. These specifications are internationally recognised and are therefore generally labelled on the sealed container in which the fluid is sold. Currently one might expect these standards to be expressed as 'SAE J1703' and 'DOT 3/4' for conventional brake fluids. A high-quality brake fluid may, of course, be formulated not only to meet but also to surpass the requirements of these specifications.

8.11 THE ENEMIES OF BRAKE FLUID

The satisfactory operation of a hydraulic braking system must ultimately depend upon the integrity of its brake fluid, which can deteriorate in the presence of air, water and dirt for the following reasons:

Air: Pure brake fluid is virtually incompressible. If it contains air then the fluid mass becomes compressible to an extent corresponding to the amount of air entrained. This results in the brakes becoming less responsive during application and is indicated by a combination of excessive travel and 'spongy' feel of the brake pedal.
Water: Brake fluid is by its nature 'hygroscopic', which simply means that it tends to absorb moisture from the surrounding air. As a result the boiling point of the fluid can be seriously lowered, especially when it is considered that a water content of 5% by weight will virtually halve the boiling point and this may occur after three years service life. The important point to appreciate is that if during severe usage of the brakes a fluid reaches its boiling point, gas pockets are formed in the top of the wheel cylinders. At worst the presence of this vapour will, between brake applications, force fluid back into the reservoir until suddenly insufficient fluid can be trapped between the master-cylinder and wheel cylinder pistons to operate the brakes at all. It is to avoid such a dangerous situation as this, that brake equipment manufacturers recommend the system fluid be changed every one year to 18 months.
Dirt: This is, of course, the enemy of all hydraulic systems where consistent operation depends upon maintaining fine clearances and effective seals. For this reason scrupulous cleanliness must be exercised when servicing hydraulic brake systems.

8.12 REMOVING AIR FROM HYDRAULIC BRAKES

The purging of air from a hydraulic brake system is commonly referred to as 'bleeding' the brakes (*Figure 8.19*). Unless the presence of air is indicated by a spongy feel of the brake pedal, bleeding should only be required after some part of the system has been disconnected, or following renewal of the brake fluid.

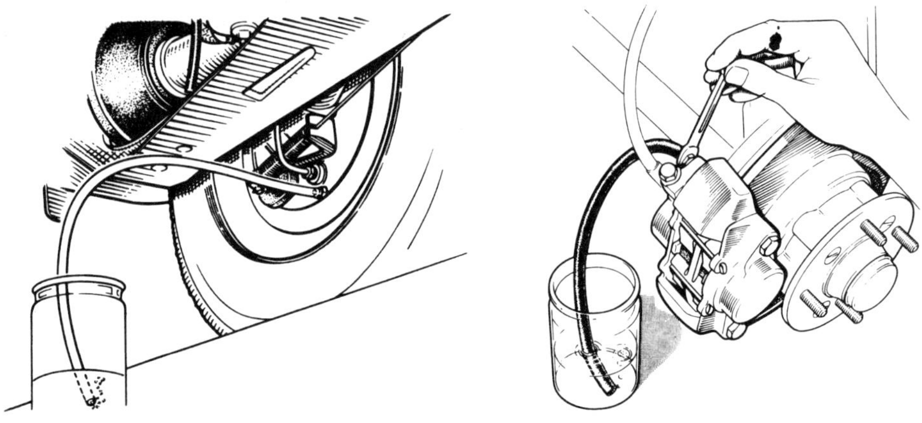

Figure 8.19 Brake bleeding (*Lockheed*)

For the purpose of bleeding the brakes a series of air bleed valves with 'bleeder screws' are employed, which communicate with the interior chambers of the various hydraulic units at their uppermost points. These units include in all cases the wheel cylinders of drum and disc brakes and in some cases the master-cylinder as well. The expulsion of air from each unit is achieved by attaching what is termed a 'bleeder tube' to the particular air bleed valve and immersing its other end in a small quantity of new fluid contained in a glass jar. If then the air bleed valve is opened a fraction of a turn and the brake system pressurised either manually by repeatedly depressing the brake pedal, or automatically by means of a commercial pressure-bleeding unit connected to the master-cylinder reservoir, any air leaving the system will be made evident by the bubbles rising in the jar of fluid. When the bubbles no longer appear the bleeder screw of the air valve is re-tightened onto its seating. The procedure is, of course, repeated at the other air bleed valves in the system.

Not so many years ago, in the era of the very basic single-line hydraulic braking system, as described in § 8.1.2 of *Vehicle Technology 1,* it would have been possible to include here a simple set of instructions that would have covered the brake bleeding procedure for practically all vehicle applications. Nowadays matters are a good deal more complicated, because from safety and other considerations the modern hydraulic braking system has reached a high level of sophistication. This in turn has led to many variations in brake bleeding procedures, which can require different pedal actions during manual bleeding, different sequences of bleeding that relate to dual-circuit safety systems, and with certain of the latter simultaneous bleeding of the front and rear brakes. Although these various requirements will be better understood when hydraulic braking systems are further discussed at a later stage, it is in any event important that the manufacturer's instructions are faithfully adhered to when carrying out brake-bleeding operations.

8.13 SERVO ASSISTANCE FOR DISC-BRAKE SYSTEMS

Sir Henry Royce is credited with having once said that 'the speed at which a car can be driven with safety is determined by its ability to stop'. It is therefore perhaps to have been expected that since the advent of four-wheel braking in the mid-nineteen-twenties, the braking system has always been one of the outstanding features of Rolls-Royce cars, combining reliable stopping power with light pedal pressures. Although such attributes are now taken for granted in the braking systems of most modern cars, this was certainly not true before the general adoption of servo-assisted disc brakes. It is therefore relevant to try and account for the superiority of the earlier Rolls-Royce braking systems. The explanation may be found first, in the use of strictly non-self-energising drum brakes at the wheels and second, in the addition of a 'servo' mechanism to multiply the effort exerted on the brake pedal by the driver. In other words, the use of low-factor wheel brakes to provide consistent braking, together with a servo mechanism to reduce what would otherwise be a high pedal effort to operate them.

Since these two basic approaches can now be found in the servo-operated disc brake systems of many modern cars, they likewise account for their much improved braking performance, or at least greatly contribute to it. The reasons for the use of a servo unit in the braking system may therefore be summarised as follows:

(1) It permits the use of low factor wheel brakes with less sensitivity to changes in lining friction as they become heated.
(2) The driver effort on the brake pedal can be reduced even further than with some high-factor, non-servo-assisted brakes.
(3) A reduction can be made in the total pedal travel required because leverage in the system can be made much less.

8.14 SUSPENDED VACUUM SERVO UNITS

These may be either of the 'indirect'- or 'direct'-acting variety. The former is actuated by hydraulic pressure from the master-cylinder and pressurises the wheel cylinders via a hydraulic slave cylinder. This type of servo unit may therefore be installed in any convenient space available. For example, twin servo units of this type are housed in a front wing compartment of the high-grade Bristol car. A direct-acting (or 'integral') servo unit is, as its description suggests, directly actuated by the brake pedal and better lends itself on most cars to the modern requirement for dual hydraulic braking systems.

In either event its purpose is to reduce the driver effort required on the brake pedal. It performs this function by exposing a servo piston to the difference between atmospheric and engine intake manifold pressures. The piston is provided with either a sliding or a rolling seal against its cylindrical housing or chamber.

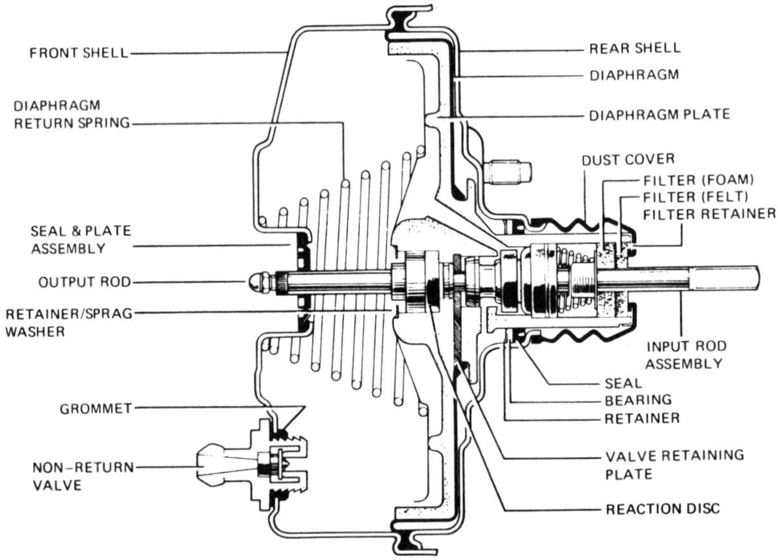

Figure 8.20 Cross-section through a direct-acting suspended vacuum brake servo (*Girling*)

If used in conjunction with two-stroke or diesel engines, the servo unit must be connected to a separate vacuum pump. With earlier designs of vacuum servo units, each side of the piston was exposed to atmospheric pressure when the brakes were released and then manifold depression was allowed to act on the leading side of the piston during their application. This was known as the 'air suspended' type of vacuum servo. In established modern practice, however, the 'vacuum suspended' type of vacuum servo is used in which each side of the piston is exposed to manifold depression when the brakes are released and then atmospheric pressure is admitted to the trailing side of the piston during their application.

The reasons for this reversal in roles are threefold, as follows:

(1) Less demand on the source of depression during brake application provides a more responsive control.
(2) Since only the trailing side of the piston need be exposed to the atmosphere there is a reduced risk of damage by the ingress of foreign matter.
(3) A simpler valve arrangement may be used requiring only one port each for the vacuum and atmospheric connections.

A 'fail-safe' feature common to all types of vacuum servo-operated braking systems is that in the event of failure of the vacuum source, the hydraulic circuit(s) remain unaffected although, of course, the driver pedal effort must be increased.

8.14.1 Operation of the suspended vacuum servo unit

Referring first to the general arrangement drawing that names the parts of a typical direct-acting servo unit and then to the simplified diagrams, the operating principle (*Figure 8.21*) may be summarised as follows:

(1) With the brakes released the servo piston is fully retracted by its return spring. Similarly the input rod connection to the brake pedal is retracted by the return spring for the latter. In this position of the input rod the vacuum port is open and there is a depression acting on both sides of the servo piston, which can thus be described as being 'suspended' in vacuum.

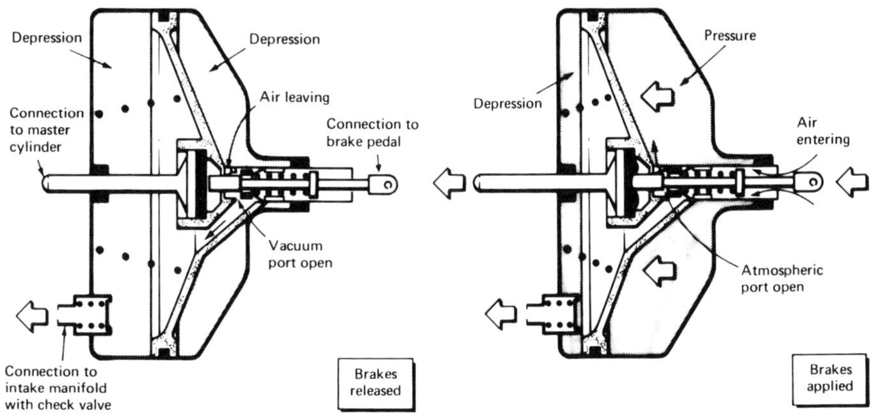

Figure 8.21 Operating principle of a direct-acting, suspended vacuum brake servo

(2) When the brakes are applied the pedal urges the input rod forward, so that first the piston control valve closes the vacuum port and then its continued movement opens the atmospheric port. Air then enters the servo unit with the result that atmospheric pressure acts behind the servo piston and assists the input rod in travelling forwards, the piston in turn actuates the output push-rod connecting to the master-cylinder plunger. During release of the brakes the air that entered the servo unit is exhausted via the opening vacuum port first to the front portion of the chamber and thence past the non-return valve to the intake manifold.

Two additional points that deserve mention are the following:

(3) When the brakes are held on, as distinct from being applied by increasing effort on the pedal, both vacuum and atmospheric ports remain closed so that a state of equilibrium exists between the master cylinder and the servo unit.

(4) The function of the rubber reaction disc is to impart a certain degree of 'feel' or 'feedback' into the pedal effort being exerted by the driver, because when this is sufficient to open the atmospheric port of the piston control valve, it also means that the valve plunger head has penetrated the reaction disc. Since the latter then acts as a return spring to eject the valve plunger, the driver is made aware of this by an opposing force felt at the brake pedal.

8.15 COMPRESSED-AIR BRAKE SYSTEMS

Commercial vehicle air-brake installations constitute a true 'power', as distinct from 'power-assisted', braking system. This is because there is no direct connection in any mechanical or hydraulic sense between the brake treadle and the wheel brake chambers, although the driver is provided with a certain degree of 'feel' related to system air pressure during braking.

In simple terms, the operation of an air-brake system is such that when the brake treadle is depressed one of two related valves is opened, so that air under pressure from the reservoir can pass through the control valve and into each wheel brake actuating chamber. Here the compressed air acts against a diaphragm, its resulting movement being transmitted via a push-rod to the operating lever of the brake camshaft, the rotation of which expands the shoes against the drum. As the brake pedal is released, the previously-mentioned control valve closes and the other one is opened, thereby allowing the air under pressure in the brake actuating chambers to be exhausted to the atmosphere and the shoe return springs to release the brakes.

Air-braking systems received their early development in America, notably by the Westinghouse Air Brake Company, and then gradually gained acceptance elsewhere from the mid-nineteen-thirties onwards. More recent years have seen the operation of commercial vehicles where both gross weight and speed capability are much greater than ever before. This in turn has led to an increasing degree of sophisticated engineering in air brake systems, not least to meet the safety directives of the EEC (European Economic Community). Indeed, it may be said that air brake technology has virtually become a separate branch of automotive engineering.

8.15.1 Layout of a basic air-brake system

At this stage we are only required to have knowledge of the layout of a basic, single-circuit, air-brake system for a four-wheeled commercial vehicle, the purpose of the various components being as follows:

Compressor: This is a reciprocating piston pump driven from either the engine or the gearbox and provides the compressed air to operate the braking system and also the vehicle ancillary services.
Reservoir: This stores the compressed air until it is needed for operating the brakes. The storage capacity must be such as to allow for several brake applications following any failure of the compressor to deliver air.
Non-return valves: Also known as 'check valves', their purpose is to allow the air to flow in one direction only, such as from the compressor to the reservoir but not vice versa, in the event of compressor failure.
Unloader valve: Mounted between the compressor and the reservoir its purpose is to relieve the working load on the former by controlling the pressure of air in the latter, usually between 665 to 840 kN/m^2.
Safety valve: This provides a safeguard against any excessive build-up of air pressure in the system and may be located either at the compressor, unloader valve, or reservoir.
Pressure regulator valve: Its purpose is to ensure that adequate air pressure exists in that portion of the system preceding the valve, before any pressure is allowed to build up in the system beyond it.

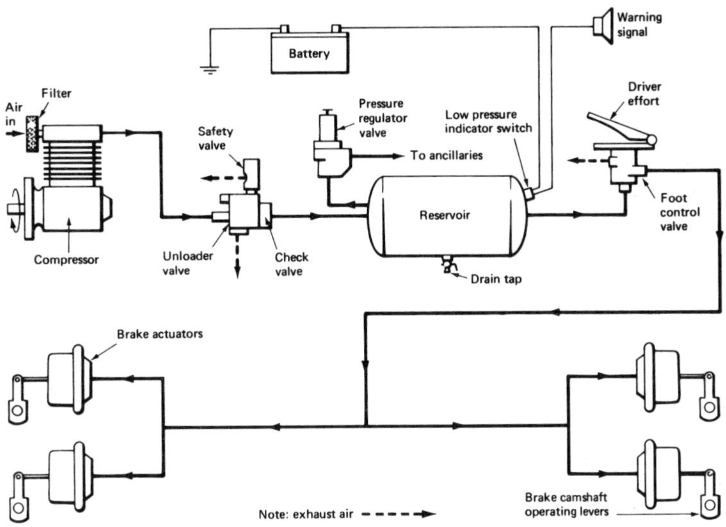

Figure 8.22 Layout of a basic single-circuit air-brake system

For example, such a valve can guarantee that a braking system reservoir receives air charging priority over an ancillary services reservoir.

Foot control valve: This is treadle operated by the driver and its function is to provide precise control of operating air pressure during application and release of the brakes. It is installed in series with the reservoir and the brake actuators.

Brake actuators: Also known as 'brake chambers', one of these is mounted externally to each wheel brake and converts the energy stored in the compressed air into the mechanical force and movement required to actuate the brake shoe expanding mechanism.

Low pressure indicator: This switch is installed in the system to give an automatic warning signal to the driver in the event of the system air pressure falling below a safe working minimum of generally 420 kN/m^2. It operates either a warning light or buzzer in the driving compartment.

8.15.2 Advantages of air-brake systems

For medium to heavy commercial vehicles these may be listed as follows:

(1) The system will tolerate a certain amount of air leakage without failing completely.
(2) Large operating forces to expand the brake shoes can readily be generated.
(3) A supply of compressed air is a convenient source of energy to operate various ancillary equipment on the vehicle.

8.16 SHOE ADJUSTMENT WITH AIR BRAKES

This is accomplished by what is known as a 'slack adjuster', the mechanism of which is embodied in the lever arm used between each brake actuator push-rod and expander cam for the shoes. By means of a lockable worm-and-wheel mechanism the rotational position of the expander cam may be manually adjusted relative to that of the lever arm. Hence, the most favourable operating geometry for the

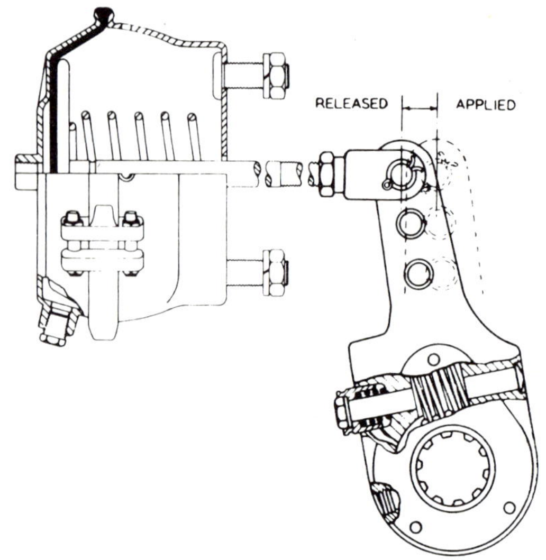

Figure 8.23 Air brake slack adjuster (*Bendix*)

actuating linkage may be preserved throughout the life of the shoe linings and by the same token the actuator diaphragm is never over-extended. During normal brake operation the entire lever arm-cum-slack adjuster does, of course, rotate bodily with the brake camshaft.